T0122371

Computational Social Sciences

Computational Social Sciences

A series of authored and edited monographs that utilize quantitative and computational methods to model, analyze and interpret large-scale social phenomena. Titles within the series contain methods and practices that test and develop theories of complex social processes through bottom-up modeling of social interactions. Of particular interest is the study of the co-evolution of modern communication technology and social behavior and norms, in connection with emerging issues such as trust, risk, security and privacy in novel socio-technical environments.

Computational Social Sciences is explicitly transdisciplinary: quantitative methods from fields such as dynamical systems, artificial intelligence, network theory, agent based modeling, and statistical mechanics are invoked and combined with state-of the-art mining and analysis of large data sets to help us understand social agents, their interactions on and offline, and the effect of these interactions at the macro level. Topics include, but are not limited to social networks and media, dynamics of opinions, cultures and conflicts, socio-technical co-evolution and social psychology. Computational Social Sciences will also publish monographs and selected edited contributions from specialized conferences and workshops specifically aimed at communicating new findings to a large transdisciplinary audience. A fundamental goal of the series is to provide a single forum within which commonalities and differences in the workings of this field may be discerned, hence leading to deeper insight and understanding.

More information about this series at http://www.springer.com/series/11784

Corinna Elsenbroich • David Anzola
Nigel Gilbert
Editors

Social Dimensions of Organised Crime

Modelling the Dynamics of Extortion Rackets

Editors
Corinna Elsenbroich
Department of Sociology
Centre for Research in Social Simulation
University of Surrey
Guildford, Surrey, UK

David Anzola
Department of Sociology
Centre for Research in Social Simulation
University of Surrey
Guildford, Surrey, UK

Nigel Gilbert
Department of Sociology
Centre for Research in Social Simulation
University of Surrey
Guildford, Surrey, UK

ISSN 2509-9574 ISSN 2509-9582 (electronic)
Computational Social Sciences
ISBN 978-3-319-83229-6 ISBN 978-3-319-45169-5 (eBook)
DOI 10.1007/978-3-319-45169-5

Printed on acid-free paper

This Springer imprint is published by Springer Nature
The registered company is Springer International Publishing AG
The registered company address is: Gewerbestrasse 11, 6330 Cham, Switzerland

To Rosaria, whose intellectual curiosity and determination made this project possible. The world has lost a great spirit.

Contents

Contributors

Giulia Andrighetto, Ph.D. Institute of Cognitive Sciences and Technologies, Italian National Research Council (CNR), Rome, Italy

Schuman Centre for Advanced Studies, European University Institute, Fiesole, Italy

David Anzola, Ph.D. Department of Sociology, Centre for Research in Social Simulation, University of Surrey, Guildford, Surrey, UK

Rosaria Conte, Ph.D. (Deceased)

Corinna Elsenbroich, Ph.D. Department of Sociology, Centre for Research in Social Simulation, University of Surrey, Guildford, Surrey, UK

Giovanni Frazzica University of Palermo, Palermo, Italy

Nigel Gilbert, Ph.D., Sc.D. Department of Sociology, Centre for Research in Social Simulation, University of Surrey, Guildford, Surrey, UK

Ulf Lotzmann Department of Computer Science, Institute for Information Systems Research, University of Koblenz-Landau, Germany

Vincenzo Militello University of Palermo, Palermo, Italy

Michael Möhring University of Koblenz-Landau, Koblenz, Germany

Luis G. Nardin, Ph.D. Institute of Cognitive Sciences and Technologies (ISTC), Italian National Research Council (CNR), Rome, Italy

Schuman Centre for Advanced Studies, European University Institute, Fiesole, Italy

Martin Neumann, Ph.D. Department of Computer Science, Institute for Information Systems Research, University of Koblenz-Landau, Koblenz, Germany

Valentina Punzo University of Palermo, Palermo, Italy

Attilio Scaglione University of Palermo, Palermo, Italy

Antonio La Spina Political Sciences, Luiss "Guido Carli" University, Rome, Italy

Áron Székely, Ph.D. Institute of Cognitive Sciences and Technologies, Italian National Research Council (CNR), Rome, Italy

Klaus G. Troitzsch Computer Science Department, Universität Koblenz-Landau, Rheinland-Pfalz, Germany

Chapter 1
Introduction

Corinna Elsenbroich, David Anzola, and Nigel Gilbert

1.1 The Project

For most of us extortion rackets will never blight our lives. This is lucky given the severe financial, psychological and social consequences extortion brings with it. It is also lucky as nobody is ever very far from extortion. A recent European review of extortion racketeering found that extortion rackets exist in every EU member state and the same holds for almost all countries in the world. Whilst ubiquitous, in most countries extortion rackets are an isolated phenomenon, isolated geographically or ethnically, or reserved for a criminal underclass of prostitution, drug dealing and gambling. In the communities where extortion takes hold it wreaks havoc, destroying livelihoods, if not lives, and undermining community cohesion. At the same time, extortion rackets can establish themselves within communities completely unknown to the rest of society. This combination of longevity and invisibility is a unique feature of extortion rackets, making them an interesting social phenomenon as well as a very-hard-to-research criminological phenomenon.

The book is the result of a project called Global Dynamics of Extortion Racket Systems (GLODERS) and funded by the FP7 programme. The GLODERS research project is directed towards development of computational models for the understanding of the dynamics of extortion racket systems (ERSs). ERSs, of which the mafia is but one example, are a global phenomenon, originating from a small number of seed locations. They cause disruption to economies by money bypassing official channels as well as by undermining investment, and societies by creating fear and distrust. As yet understanding of ERS is relatively poor,

C. Elsenbroich, Ph.D. (✉) • D. Anzola, Ph.D. • N. Gilbert, Ph.D., Sc.D.
Department of Sociology, Centre for Research in Social Simulation, University of Surrey, Guildford, Surrey GU2 7XH, UK
e-mail: c.elsenbroich@surrey.ac.uk; d.anzola@surrey.ac.uk; david.anzola@gmail.com; n.gilbert@surrey.ac.uk

C. Elsenbroich et al. (eds.), *Social Dimensions of Organised Crime*, Computational Social Sciences, DOI 10.1007/978-3-319-45169-5_1

resulting from the clandestine nature of the phenomenon as well as a lack of integrative approaches (Di Gennaro & La Spina, 2016). ERSs are not only powerful criminal organisations, operating at several hierarchical levels, but also prosperous economic enterprises and socially situated dynamic systems. The approach taken in GLODERS is particularly focussed on the needs of stakeholder, such as legislators and law enforcers, to further their understanding of ERS and how to tackle them. Through focussing on social dynamics and stakeholder needs, GLODERS presents a novel approach to the understanding of extortion rackets.

This book presents the results from the GLODERS project. It provides the synthesis of novel data and innovative analyses of data, a unique consortium of stakeholders and a new methodological approach to extortion racketeering, to understand the internal dynamics of extorting criminal organisations as well as extortion rackets embedded in society.

The new data consist of a unique database of extortion cases in Sicily and Southern Italy, police and court data for extortion racketeering in Italy, the Netherlands and Germany and novel uses of datasets such as the European Value Survey.

Some of these data would not have been available without the strong stakeholder involvement in the project. A stakeholder board consisting of about 30 international experts on extortion rackets was involved from the start and throughout the project. Stakeholders provided access to data not otherwise available but more importantly stakeholders provided access to expertise that was essential for the success of the project. The expertise involved narrative evidence, judgement about findings of the research and a focus on producing research focussed on real-world applications.

This kind of expertise is particularly important for the novel methodological stance explored in this book: trying to understand extortion rackets from a complex systems perspective which takes account of their embeddedness in society and the emergence of norms and behaviours within and surrounding extortion racketeering. The methods used to understand extortion racketeering are computer based and consist firstly of the simulation method of agent-based modelling and secondly of enhanced computational analysis and integration of large amounts of a variety judicial data.

Through this unique combination of data, stakeholder expertise and computational modelling there are several interesting results regarding the dynamics of extortion rackets. The most important results are the role of trust in an extortion racket, the role of social norms in sustaining as well as fighting extortion rackets and the role of civic organisations in supporting norm change in society.

1.2 Researching Extortion Rackets

Extortion rackets have been researched from multiple angles and a range of disciplines. There are several assessments of the damage to national economies caused by extortion. The damage comes about in various ways, as loss of tax to governments, loss of reinvestment into businesses as well as undermining of external investment into an extorted territory.

Schelling (1967) provides one of the earliest accounts of racketeering focussing on the interplay of the upper and the underworld. He points at three important aspects of organised crime: (a) the economic dimension, i.e. the impact of illegal markets on the economy and tax revenue of a country; (b) policing organised crime; (c) the structure of organised crime; and (d) its entrenchment in society. Although the focus is more on the black market economy than extortion rackets points (c) and (d) pertain directly to the content of this book. Part IV presents a simulation of a network of organised crime over time to its demise. The simulation shows in particular how the erosion of norms of trust leads to the collapse of the organisation.

"The underworld seems to need institutions, conventions, traditions, and recognisable standard practices much like the upper-world of business" (Schelling, 1967, p. 68).

Part III delves into analyses of the societal dimensions of extortion racketeering, the interaction of the upper and the underworld. But rather than focussing on the economy (e.g. Asmundo and Lisciandra (2008), Frazzica, Lisciandra, Punzo, and Scaglione (2016)), the research here focuses on normative aspects and the ways of changing social norms to fight and support policing of extortion racketeering.

Another large area of research into extortion racketeering is conducted using a game theoretic approach. The interaction between an extorter and a victim is a classic interdependent choice problem. Gambetta (1994) follows the general interdependent choice framework with a focus on trust and the flipside of trust: the credibility of threat and the reach of reputation. An extortion racket does not function if the racketeers have to cause a lot of violence; in fact, the low rate of violence is a hallmark of a functioning extortion racket, cf. La Spina (2008).

Formal game theory has been applied to various configurations of actors in extortion racketeering. The extortion situation intuitively reads as an iterative decision tree. The extorter demands money from an entrepreneur, the entrepreneur decides to pay or not to pay and then the extorter decides whether to punish. The demand and decision occur multiple times between a paired extorter and entrepreneur. In the game theoretic literature on extortion rackets (Gambetta, 1988, 1994; Konrad & Skaperdas, 1998; Varese, 1994, 2001), the payment decision *depends on the entrepreneur's expectation of being punished for refusal to pay*. Konrad and Skaperdas (1998) treat the probability of punishment as endogenous, arising from the efforts invested by extorters and police. The game described in Konrad and Skaperdas (1998) formalises the interdependent choice between three actors: the extorter, the entrepreneur and the police. On the other hand Gambetta (1994) and Smith and Varese (2001) have levels of police presence as an exogenous variable, focussing instead on the extorter-victim interaction. The game originates with Gambetta (1988) who uses it as a setting to discuss reputation, signalling and piracy of symbols. Extorters are split into two types, the real Mafiosi and fakers. Mafiosi establish a reputation for punishing resistance and fakers can free-ride on this reputation if they manage to signal belonging to the real mafia. Entrepreneurs decide whether to pay or not depending on their individual expected punishment probability calculated from whether they think an extorter is a faker or not, given knowledge of the level of police presence. High police presence lowers the probability of a Mafioso to punish and also the probability of fakers to extort. The game has been formally analysed in detail in Smith and Varese (2001).

The above approaches are highly theoretical or focussing exclusively on estimated economic impact. There are also more empirical approaches to the investigation into extortion racketeering.

One important source of data is in-depth interviews with victims, *pentiti*, Mafiosi that provided evidence in trials and left the organisation, and officials. Two major examples relying on these data are Varese (2001) on the Russian and Paoli (2003) on the Italian mafia.

Varese (2001) highlights the features of a transition economy on the success of protection racketeering. The book integrates a variety of data sources, from interviews with victims, mafia member and officials, over data from undercover police investigations to archival documents.

Paoli (2003) provides a close analysis of the structure of the Italian mafia, in particular on the Cosa Nostra and 'Ndrangheta. The focus is on the organisational structures, the cultural and ideological aspects of the mafia and their integration into the social and political context of Italy. The argumentation in the book partly follows that of Gambetta (1994) in identifying extortion as the essential economic MTO activity and in seeing trust bonds and family ties as being at the heart of the Italian mafia.

Quantitative approaches include several large-scale surveys of entrepreneurs and businesses, often conducted by third-sector institutions (cf. Di Gennaro and LaSpina (2016). These surveys show relatively low levels of extortion as well as intimidation in Italy. However, the surveys all have a serious problem for the assessment of extortion racketeering given a low response rate and a strong suggestion that the self-selected sample of respondents are already those not extorted (Di Gennaro & La Spina, 2016).

In addition to the descriptive research on extortion there is research on the legal aspects of extortion racketeering, their effectiveness of undermining extortion by supporting victims, law enforcement and judiciary. Much of the research is concerned with comparative analyses of legislation in EU member states. Barriers to unified EU legislation are a general focus on organised crime but very different instantiations of this kind of crime in the different memberstates. Whilst all countries are affected by drug, weapon and human trafficking, the particular MTO practice of extortion varies greatly between EU countries. Finding laws that agree with the general legal systems of countries and capture the specific needs of all member states seems elusive, although signs of slow convergence towards a civil law approach can be identified (Calderoni, 2010).

Criminology in general has a problem with procurement of adequate data; however, often the victim side is fairly reliable. For extortion racketeering also the victim response needs to be critically examined due to the long-term relationship between the extorter and the victim. For example in the case of the Italian mafia, extorted entrepreneurs might reasonably not respond to the survey as they are afraid of disclosure or colluding with the mafia (La Spina, 2008).

The problem of reliable data and other information to investigate extortion is discussed extensively in a special issue of Global Crime (2016, Vol 17, Issue 1) and is one of the motivations for the computational approach discussed in this book. Questions about the reliability of every single data source lead to a demand for the triangulation and integration of multiple data and the use of multiple methods to put together the jigsaw that is extortion racketeering.

1.3 Enhancing Understanding

As the above section suggests, integration of many methodological approaches and various data is necessary to enhance the understanding of extortion racketeering as a criminal as well as a social phenomenon. This book details a computational modelling approach to integrate several of the separate research strands discussed above and additional ones, such as participatory co-production, and large-scale European survey data.

Part I contains a comparative analysis of extortion racketeering as a global phenomenon. Chapter 2 provides information about the most prevalent global extortion rackets, including Russia, Latin America and the Yakuza, but excluding the Italian mafia. As the Italian mafia is the case study for a socially embedded extortion racket, a lot of detail is provided in Part II of the book. Chapter 3 builds on the comparative analysis of Chap. 2, extrapolating a typology of extortion rackets in societies. The typology considers the three dimensions of the structure of the criminal organisation, the civil society it is situated in and the state, including agents of the state such as law enforcement and judiciary. The typology highlights how important it is for the latter two features for extortion rackets to flourish. GLODERS focussed on two aspects of the typology, which will be extrapolated upon in two case studies in the following chapters of the book. Parts II and III focus on the interaction of the state and civic society. They look in particular at state responsiveness in the form of law enforcement as well as aspects of dissociation from the state and structural social capital, in the form of denunciation and the role of norm change towards pizzo payment.

Part II constitutes a theoretical preparation for the investigation of two aspects of extortion rackets highlighted in the typology in Chap. 3, the interaction between the state and civil society. Chapter 4 provides a close analysis of the importance of social and legal norms and their interactions for the understanding of extortion racketeering and Chap. 5 details anti-mafia legislation in Italy and the EU, thus focussing on a particular state aspect of extortion.

Part III presents the case study of the Italian mafia, detailing a new empirical basis for research and two simulation models of the interaction between state and civil society in extortion rackets. Chapter 6 provides a general analysis of the Italian mafia's extortion practices, including law enforcement and social response aspects. A unique data source for the investigation of extortion rackets is presented in the form of a database of extortion cases in Sicily and Calabria. Chapter 6 sets the stage for the models discussed in Chaps. 7 and 8. Both models represent the extortion situation in Sicily, to investigate the interplay of social norms regarding pizzo payment, state interventions and civic organisations (e.g. *addiopizzo*) on levels of extortion. The difference between the models is that in Chap. 7 parameterisations were deliberately developed, according to historical records in Italy, whereas in Chap. 8 a Monte Carlo parameterisation was tested in order to find out a large variety of simulation outcomes. Chapter 9 brings together several data sources to help validate and interpret the models.

Part IV investigates a second aspect of the typology developed in Chap. 3: the criminal organisation itself. The analysis is executed on a second case study, investigating the relationships and actions of agents in a criminal network engaged in extortion practices. Chapter 10 lays the data foundation for the simulation, applying a conceptual modelling tool (CCD) to transcripts of police interrogations of suspects and witnesses. The conceptual model is translated into a simulation replicating the process of the criminal organisation's breakup in Chap. 11.

Part V provides a synthesis discussing general aspects of simulation modelling, in particular focussing on questions of validation and the integration of different kinds of data in Chap. 12. Chapter 13 provides a general conclusion of the book, bringing together the theoretical, methodological and empirical aspects of investigating extortion racketeering.

References

Asmundo, A., & Lisciandra, M. (2008). The cost of protection racket in Sicily. *Global Crime, 9*(3), 221–240.

Calderoni, F. (2010). *Organized crime legislation in the European Union: Harmonization and approximation of criminal law, national legislations and the EU framework decision on the fight against organized crime*. New York: Springer.

Di Gennaro, G., & La Spina, A. (2016). The costs of illegality: A research programme. *Global Crime, 17*(1), 1–20.

Frazzica, G., Lisciandra, M., Punzo, V., & Scaglione, A. (2016). The Camorra and protection rackets: The cost to business. *Global Crime, 17*(1), 48–59.

Gambetta, D. (1994) Inscrutable Markets, Rationality and Society, 6, 353-368.

La Spina, A. (2008). *I costi dell'illegalità. Mafia ed estorsioni in Sicilia*. Bologna: Mulino.

Konrad, K. and Skaperdas, S. (1998) Extortion, Economica, 65:260, 461–477.

Paoli, L. (2003). *Mafia brotherhoods: Organized crime, Italian style*. New York: Oxford University Press.

Schelling, T. (1967) Economics and Criminal Enterprise, Public Interest, 7.

Gambetta, D. (1988) Fragments of an Economic Theory of the Mafia, European Journal of Sociology, 29:1, 127–145.

Varese, F. (1994) Is Sicily the future of Russia? Private protection and the rise of the Russian Mafia, European Journal of Sociology, 35, 224–258.

Varese, F. (2001). *The Russian Mafia: Private protection in a new market economy*. Buckingham: Open University Press.

Part I
Extortion Rackets as a Global Phenomenon

Chapter 2
National Mafia-Type Organisations: Local Threat, Global Reach

David Anzola, Martin Neumann, Michael Möhring, and Klaus G. Troitzsch

2.1 Introduction

A first challenge in characterising and quantifying the nature and dimension of the extortion rackets phenomenon worldwide is that extortion is an umbrella concept grouping a large array of criminal practices. Extortion is often associated with long-standing and well-organised criminal organisations, such as the Italian mafia and the Japanese Yakuza, given the amount of data and research about these criminal groups and their popularity in contemporary popular culture. Yet, a review of the different extortion practices around the world quickly makes readily available the significant diversity and complexity of the social contexts in which extortion occurs. This chapter provides a brief review of typical dynamics of extortion in different countries around the world. The main goal is, first, to summarily show the persistence and diversity of extortion rackets worldwide and, second, and most important, to contextualise the two cases analysed in the following chapters, by providing some points of contrast regarding the social conditions of the phenomenon of extortion, as well as their academic accessibility.

D. Anzola (✉)
Department of Sociology, Centre for Research in Social Simulation,
University of Surrey, Guildford, Surrey GU2 7XH, UK
e-mail: d.anzola@surrey.ac.uk

M. Neumann
Department of Computer Science, Institute for Information Systems Research, University of Koblenz-Landau, Universitätsstr 1, Koblenz 56070, Germany

M. Möhring
University of Koblenz-Landau, Koblenz, Germany

K.G. Troitzsch
Computer Science Department, Universität Koblenz-Landau, Universitätsstraße 1, Koblenz, Rheinland-Pfalz 56070, Germany
e-mail: kgt@uni-koblenz.de

C. Elsenbroich et al. (eds.), *Social Dimensions of Organised Crime*,
Computational Social Sciences, DOI 10.1007/978-3-319-45169-5_2

Four main cases are analysed through the text. The first one is the Japanese Yakuza, a very popular and thoroughly studied criminal organisation. Extortion is extremely common in Japan. The Japanese context is distinctive, for there is strong social and political legitimation of several extortive practices: first, because of the long-standing bonds between criminal organisations and the state and right-wing elites and, second, because of the perceived relative advantage civilians have in consuming some of the services or goods provided by criminal groups in the form of extortion.

The second case is the Russian mafia. Along with the Sicilian mafia and Japanese Yakuza, the Russian mafia is probably one of the most well-known criminal organisations in popular culture. They all have a similar context of origin, i.e. important social and political transitions, and have achieved similar level of penetration of social, economic and political life. Yet, contrary to its Japanese and Italian counterparts, extortion practices of the Russian mafia are not well documented. Information is scarce and unreliable. This is partly because of operational and methodological difficulties in the collection of data, but also because extortion is often subordinate of other more important types of crimes committed by these criminal groups.

The third case is Latin America. This region provides an interesting mixture of criminal organisations engaging in different forms of extortion. None of these groups, however, fits entirely into the mold of a mafia-type organisation (MTO). In spite of lacking the level of institutional penetration of more well-known mafia groups, some types of extortion thrive in the region because of widespread conditions of violence, exclusion and deprivation and the weak reliance on geographic factors, such as exclusive control of the territory, of some common extortive practices in the region.

Finally, the last case discussed is Germany. Unlike some of the other countries described, most extortion in Germany is performed by criminal organisations that did not originate in the country, such as the Sicilian or Russian mafia, or motorcycle gangs, which are often chapters of large motorcycle gangs with worldwide presence. Extortion in Germany is not strongly linked to a historically advantageous institutional framework for criminal groups or impoverished social conditions. It does not seem to constitute a serious threat to citizen security, either. Still, the case is interesting because it shows the international reach of some criminal organisations and the challenge for governments and research organisations, in terms of developing adequate tools to measure the impact of the transnationalisation of extortion practices.

Before diving into the analysis, it is important to set a couple of distinctions. Extortion is often classified in terms of its extension in time and the nature of the victim-perpetrator relationship (TRANSCRIME, 2008). Regarding the temporal extension, there is a differentiation between casual and systemic extortion. The former is an one-off episode, whereas the latter involves a relationship that extends over time. This difference is important in the present context, for most extortion carried out by large criminal organisations is somewhat systemic, given the institutional penetration and amount of resources these organisations have at their disposal. Regarding the victim-perpetrator link, the relationship, following a biological analogy, can be classified as predatory, parasitic and symbiotic. The first one implies a casual exploitation of the victim; the second, a systemic relationship of exploitation; and the third, a systemic relationship in which both parties receive some benefit from the interaction. Most criminal organisations tend to combine more than one type of extortion of this second category. The type chosen usually gives important clues about the level of institutional penetration achieved by the criminal organisation, the amount of resources invested by the victim and perpetrator and the level of social or legal legitimisation of the interaction.

2.2 Yakuza

The Japanese Yakuza is one of the oldest ERSs with worldwide reach. "Yakuza" is an umbrella concept, grouping several criminal syndicates, originated in Japan since the early twentieth century. These criminal organisations are well entrenched in social institutions, developing several mechanisms of control of social life, including extortion. These control mechanisms are strengthened and validated by historically resilient links with the Japanese Government and the right-wing elite. Kaplan and Dubro (2003) quote a Japanese social critic as saying: "Extortion is to Japan as snow is to the Eskimos. There are a hundred variations" (p. 158). These variations can be generally grouped in casual and systemic, although it could be argued that most extortion methods in Japan is, in fact, somewhat systemic.

2.2.1 Systemic Extortion

Protection rackets are probably the most common mode of systemic extortion. Many of these rackets occur in illegal industries, tolerated both by the authorities and the Japanese society, because of their relative innocuous character. Japan, for example, has the largest sex market for women in Asia, producing annual profits between ¥4 and ¥10 trillion, which represents between 2 and 3 % of the country's GNP (Dean, 2008). As of 2006, there were 1200 brothels and 17,500 sex-related businesses, such as massage parlours and strip clubs (Hongo, 2008); all this in spite of the fact that prostitution is illegal in Japan. The Yakuza have taken advantage of the absence or ambiguity of the legal framework and developed several extortion rackets focused on preventing the disruption of business and providing quick conflict resolution. These services are also widely sought for by legal establishments, such as clubs, bars and restaurants. A 1995 police survey of entertainment businesses in Tokyo revealed that almost one-third of the 60,000 establishments surveyed were paying protection money (Kaplan & Dubro, 2003).

Extortion rackets are also common in labour-intensive industries, such as construction. By the late 1990s, up to 50 % of public construction projects in Japan paid extortion money to the Yakuza, ranging between 2 and 5 % of the total construction cost (Hill, 2006). While the anti-mafia measures of the last couple of decades have cracked down hard on Yakuza-linked construction companies, these criminal organisations still profit from this business, thanks to the use of extortion rackets. Yakuza extortion rackets in the construction business cover two basic aspects. The first one is labour. The sector depends on large flows of unqualified workers. Informal labour brokering takes a staggering 70–80 % of the labour exchange market in Japan (Hill, 2006). Yakuza influences these flows through labour brokerage. Rackets are also used to control the everyday operation of construction. The labour-intensive construction sector is very susceptible to delay and sabotage. Along with the provision of sufficient and well-behaved workforce, Yakuza offers protection for things such as theft or damage of machinery and construction materials.

Extortion rackets in corporate Japan are not limited to labour-intensive industries. For years, the companies listed on the stock market have been targeted by a distinctive type of financial racketeering known as *sōkaiya*. This form of financial racketeering was developed by criminals outside the Yakuza. Yet, these criminal syndicates quickly took notice of the profit generated by this extortion method and started taking over the *sōkaiya* business during the 1970s. By the early 1980s, there were around 6800 men working on this type of extortion, distributed in over 500 separate groups and extorting as much as $400 million a year (Kaplan & Dubro, 2003).

The principle behind *sōkaiya* is relatively simple. Criminals would buy shares of the targeted company in the stock market, which grants them permission to attend the annual shareholder meeting. Once there, they extort the company with threats as simple as disrupting the meeting. Most *sōkaiya* have more sophisticated threats, however. It is common for criminals to show up at the meeting and present the directors with scandalous or embarrassing information about them or the company, e.g. irregular payoffs or bookkeeping, safety issues and mistresses. A 1999 survey showed that Japanese companies were paying these regular extorters an average of $2000 a year, and double that amount for a selected few "expert" *sōkaiya* (Kaplan & Dubro, 2003).

The practice of *sōkaiya* changed when organisations started hiring their own *sōkaiya*, either to protect them against other *sōkaiya* or as private security forces. A large retail company, for example, paid a Yakuza-linked *sōkaiya* ¥160 million to keep shareholder meetings in order between 1994 and 1995 (Hill, 2006). This change in extortion practices has eventually led to the distinction between in-house, *yotō-sōkaiya*, and outsider, *yatō-sōkaiya*. The former provides a protection service that is symbiotic; the second attempts predatory or parasitic extortion. Yakuza syndicates are often at both sides or the practice.

2.2.2 Casual Extortion

The Yakuza also engages in different kinds of extortion beyond the traditional protection rackets. Casual one-off types of extortion target both individuals and companies, and extort money or specific goods provided by the corporations. In some cases, casual extortion is partly associated with extortion rackets.

Criminal organisations in Japan are powerful enough to extort the largest and more important companies. Along systemic modes such as *sōkaiya*, there is quite a diverse variety of casual types of corporate-based extortion. A 1991 police survey of 3000 large firms in Japan found that around 41 % had been extorted by the Yakuza (Rankin, 2012). These companies are extorted into giving positions, money or different kind of goods or services to the criminals. In some cases, these extortions are done in order to further increase the criminals' reach in legal businesses. During the early 1990s, for example, the DKB bank was extorted into lending ¥26 billion to a well-known *sōkaiya*. This money was then used by the criminal to buy shares of a company he later extorted (Hill, 2006).

Some of these modes of casual extortion have moved from the parasitic to the predatory. The Yakuza, for example, has historically intervened in processes of bankruptcy management. Criminals were often hired by both the company management and the creditors. The former would hire them to provide protection, and the latter, for debt collection. This privileged position would allow the Yakuza to manipulate the whole process. In occasions, this ability to manipulate the entire negotiation process would lead creditors to sell their debt to the Yakuza, sometimes for as little as 5 % of face value (Hill, 2006).

These extortion schemes were eventually taken further. Extorters identify companies in the break of bankruptcy, for which they provide short-term financial support. From there, they force the manager to give them enough access and power in order to advance their position as creditors. Once the company finally goes bankrupt, and sometimes even before that, the Yakuza takes over its assets. It is difficult to estimate the extent of this kind of extortion, but it should be significant, taking into account that, through the 1980s, an average of 18,000 companies, with debts over ¥3.6 trillion were declared bankrupt annually (Hill, 2006).

Beyond the corporate world, the most important mode of casual extortion is associated with the provision of conflict resolution. The prevalence of this mode of extortion is not so much linked to Yakuza's power, as it is to the inefficiency and limitations of Japanese bureaucracy and legal system. Civilians prefer the intervention of the Yakuza, in order to avoid a system that is both slow and expensive. Debt collection, for example, can only be performed by lawyers. When done legally, the process is often extremely slow, so lenders prefer to pay a hefty fee to the Yakuza, so as to speed up the process. Criminals usually keep 50 % of the debt, plus expenses, which, in some cases, could leave the lender with as little as 20–30 % of the original debt (Hill, 2006). A similar situation occurs with individuals looking forward to resolving traffic disputes, businessmen looking to get official permits processed and ordinary citizens seeking to get a hold of official authorities. They are all forced to rely on criminal organisations to get things done quickly.

The Yakuza have also developed a particular scheme of casual extortion known as *pretext extortion or racketeering*. The criminal basically claims compensation for a good or service for which a relatively trivial fault is found (Hill, 2014), for example, a bug in the food, a purchased item of substandard quality or a service or good that is different than advertised. Pretext racketeering is sometimes carried out by creating fake social and political movements or by infiltrating existing ones, which makes the extortion escalate both in terms of complexity and profit. A 1986 survey of 5030 companies in 17 different industries found that around 26 % of them had been extorted by social groups protesting on discrimination-based claims (Hill, 2006).

The Yakuza is naturally present in traditional areas of predatory extortion, such as loan-sharking. This activity is particularly profitable in Japan, since the formal financial system is not entirely integrated and financial services are extremely fragmented. As much as 10 % of the Japanese population resorts on moneylenders, which lend money under huge interest rates (The Economist, 2008). By 1983, it was 73 % a year. This value has constantly been reduced to a low of 20 % in 2006. The Yakuza have been increasingly pushed out of the legal moneylending business, but they still operate a parallel illegal system in which money is lent, for example, adding 10 % of the debt every 10 days (Kaplan & Dubro, 2003).

The landscape of casual extortion carried out by the Yakuza varies with contextual conditions. Due to the real estate bubble of the 1980s, for example, the Yakuza entered into the business of land-sharking. The goal of this type of extortion was to force landowners or lease holders to sell or give up their lease, respectively, so as to allow larger real estate developments on that land. This extortion was usually paid by big developers, which were usually charged 3 % of the land value by the criminals (Hill, 2006). This was an extremely popular practice during the 1980s. By 1990, the Osaka police had documented up to 1600 land-sharking denunciations (Kaplan & Dubro, 2003). It, however, lost its attractiveness after the real estate bubble burst, which significantly decreased land prices for more than 20 years. By 2008, property prices in Japan were only about 40 % of their values before the start of the crisis (McCurry, 2008).

The widespread character of extortion rackets in Japan is the consequence of both the inefficiency and complacency of local authorities. Regarding the latter, historically, there has been a strong link between the Yakuza, the Japanese Government and right-wing elites. This connection has allowed the Yakuza, first, to take over important economic sectors, such as construction, and, second, to gain important leverage when it comes to issuing policy aiming at crackdown on organised crime. The inefficiency of the Japanese Government has, on the other side, allowed the Yakuza to position themselves as crucial brokers of social life. The Yakuza has traditionally focused on the provision of social and public services and goods. While the provision of these goods and services is usually underlain by an unfair advantage for the criminals, many victims actively seek for the Yakuza services, since they are in different ways perceived to be better than official channels.

2.3 Russia

"Russian mafia" is a generic way to refer to a vast array of criminal organisations that emerged within the territories of the former Soviet Union. These organisations developed criminal enterprises strongly linked to the communist system and gained significant power during the transition to capitalism. Like the *Yakuza* and *Cosa Nostra*, there are conflicting foundational myths regarding the origin of these criminal groups. Sometimes their origin is traced back to criminal organisations that emerged in the early twentieth century, during tsarist times. Yet, most accounts trace the origin of these groups to the *Vory v zakone* (thieves-in-law) of the Soviet *gulags*. These groups allegedly emerged as an informal mechanism of control of the prisoners in the *gulags*, but quickly became highly organised criminal groups that took full advantage of this level of organisation once the *gulags* were closed and the prisoners were out.

The different criminal groups that compose the Russian mafia are engaged in numerous types of crimes, ranging from petty crime to highly coordinated transnational criminal activities, e.g. drug, human and gun trafficking. This diversity in the criminal activities in which these groups take part is, to a certain extent, an effect of their high structural flexibility. The best detailed overview of different criminal organisations around the world—albeit a little outdated—can be found in the "Violent Non-State Actors Database", compiled by the *Matthew B. Ridgway Center for International Security Studies*.[1] This database contains information for the six main Russian mafia-like brotherhoods currently operating worldwide: The Izmailovskaya, the Mazukinskaya, the Liuberetskaya, the Podolskaya, the Solntsevskaya and the Tambovskaya. Except from the last one, all these criminal organisations were named after suburbs or city districts in the Moscow region. They all were founded between 1985 and the early 1990s. Structurally, they all adopt a vertical structure, but are heavily fragmented in networks spreading all over the world. Central coordination diffuses as the cells are farther away from central in the Russian Federation (for an older source see Volkov, 2002).

In terms of extortion, the Russian mafia has followed a path quite similar to that of the Yakuza and the Sicilian mafia. It amassed significant power through the provision of protection rackets during the tumultuous period of transition between socialism and capitalism. According to Gans-Morse (2012), after the collapse of the Soviet Union, "[c]ourts, law enforcement bodies, and state regulatory agencies capable of enforcing the rules of the game for a modern market economy had to be created from scratch or rebuilt from the remnants of socialist institutions" (p. 263). Because of the lack of adequate protection and the importance of property rights in a market economy, several private actors turned to criminal organisation for the provision of protection of the recently acquired or expanded property rights. This role was not limited exclusively to protection, but also covered some other aspects related to the everyday economic transactions in a market economy. "In the absence of effective state institutions, firms turned to alternative forms of protecting property and enforcing contracts. Criminal protection rackets and private security agencies provided physical protection, collected debts, and adjudicated disputes among firms" (p. 263).

While protection became a particularly sought-out service during the transition period, the overall brokering of social life was a role criminal organisations were regularly engaged with way before the collapse of the Soviet Union.

[1] http://research.ridgway.pitt.edu/topics/gang-intro/violent-non-state-actors-database/, last opened September 8, 2015 11:58.

According to Burton and Burges (2007), criminal organisations were often involved in the provision of social goods and services for "Corruption in the Soviet Union was bred largely by a state-run economy that left citizens lacking basic goods. Small groups of entrepreneurs emerged to provide items otherwise not available—and the black market was born". Reports about the dimension of the second economy of the Soviet Union tend to agree that it experienced a rapid increase between the 1960s and the 1980s and that, right before the collapse of the Soviet Union, it had a large participation in the nation's GDP (Alexeev, 1995). These dynamics had important ripple effects in post-Soviet economies for years to come. Around 1998–1999, Georgia's shadow economy was still significantly large, at 64 % of the country's GDP (Schneider & Enste, 2002).

The role of social brokers of criminal organisations was not significantly altered with the transition to capitalism.

> Since the Iron Curtain fell, Russian organized crime groups have used the economic reforms and crises to increase their wealth and influence. [...] Over the years, organized crime also has conducted certain functions of the government, including dividing territory among competing economic actors, regulating business markets, imposing 'taxes' (protection fees) and setting up tariffs, legitimizing the mafia in the eyes of many Russians as a type of de facto government. (Burton & Burges, 2007. This is also argued in Hignett, 2012)

Interestingly, however, after Putin's rise to power, the Russian Government has taken away from the Russian mafia some of its traditional extortion markets (Volkov, 2014), because of the country's transition into a competitive authoritarianism, in which the government focuses on the accumulation of institutional power under the central government, but without doing away with democratic institutions (Levitsky & Way, 2010). In order to succeed in the implementation of this political system, the Russian Government needed to get rid of other major institutional actors, including the mafia.

The relative governmental success in pushing the mafia away from its traditional zone of influence is due to the fact that, unlike other traditional mafia-type organisations, Russian criminals did not care much for the development of widespread trust networks within the local population. As a result, a significant part of the population that depended on the mafia rackets started using legal institutions and mechanisms, following the government crackdown on the mafia (Taylor, 2011). The Russian mafia lost a hold of these people, even though, as of 2012, 64 % of the Russians claimed that they usually try to go along with their business without involving governmental institutions and 63 % reported having a negative image of the law enforcement institutions (Levada Analytical Center, 2013).

While the overall dynamics of extortion rackets in Russia and other former Soviet states where the Russian mafia has strong presence has been documented, precise quantification of the phenomenon has not been possible. There is far less information about this ERS than about any other major criminal organisation. Russian official sources are more or less taciturn about this phenomenon. Internet sources—however reliable they might be—rarely go beyond about 2005. Potential Russian stakeholders who promised to contribute to GLODERS in the early phase of the project withdrew for mostly unknown reasons. One can only speculate about the reasons of this scarcity of reliable information: either it is the reluctance of

Russian academics to do research into this type of crime or the desire of Russian official administrative and statistical bodies to leave this type of crime in the dark—or, as Gans-Morse (2012) believes, the prevalence of extortion has decreased in Russia for several years, at least for the criminal racketeering, and been replaced by "state threats to property rights. [...] Key threats include seizing firms' assets, illegal corporate raiding, extortion, illicit fines, and unlawful arrests of businesspeople" (p. 263). Hence, according to this source, the influence of mafia-like groups was replaced by official administrative behaviour of a similar severity.[2]

2.4 Latin America

There are several ERSs currently operating across Latin America. Criminal organisations such as drug cartels, guerrilla movements and local and international gang are all involved in different types of extortion practices. These ERSs differ in structure, mode of operation and goals, but share a relatively homogeneous context of origin and operation that accounts for their most distinctive features. The emergence and propagation of diverse ERSs in Latin America are strongly linked to a historically generalised context of exclusion, deprivation and violence. By 2014, there were 96 million poor and 71 million indigents, most of them indigenous, black and/or living in rural areas. In turn, while the richest decile controls around 40 % of the total income, the four poorest control only 10 % and the first seven only around 30 % (ECLAC, 2014). Regarding violence, rates of violent crimes in Latin America are sixfold the numbers for the rest of the world. Homicide is particularly critical. The region accounts for only 8 % of the world's population, but 40 % of the homicides (Felbab-Brown, 2011). Firearms are used in around seven out of every ten homicides in the region, while in the rest of the world it is around four out of every ten (UNODC, 2013). Organised crime has a lot to do with these numbers. Organised crime/gang-related homicide accounts for around 30 % of the homicides in Latin America, compared to less than 1 % in Asia, Europe and Australia (UNODC, 2013).

ERSs in Latin America differ from traditional mafia-like organisations in their connection to territory. Territorial control is often considered a basic prerequisite of extortion (Block, 1980; Paoli, 2003). Yet, in the region, territorial control is a subordinate resource. It is still possible to find traditional protection rackets. Some of these rackets, in fact, have achieved a high degree of sophistication and formalisation. In some regions of Colombia, for example, the FARC issues no debt certifications, in order to allow for a smother operation of their extortion rackets (El Espectador, 2014). Yet, unlike traditional MTO, extortion practices in Latin America are usually more aligned with the needs of the drug business, which relies on local

[2] Otherwise, recent articles in Post-Soviet Affairs touch the mafia problem only superficially. "Extortion" has never been mentioned in this journal since 1999.

networks, civilians and strategic alliance between major criminal organisations (Chalk, 2011; UNODCCP, 2000).

Extortion in the region does not depend on territorial sovereignty either. The FARC, for example, have developed long-standing protection rackets over the drug production and transportation in Colombia (McDermott, 2014). They, however, do not control the business or the criminal organisations behind it directly (Otis, 2014). Non-sovereign protection rackets are common in Latin America, probably due to the multiplicity of actors in the region. These rackets are also underlain by strategic political or economic alliances. The Zetas, for example, provide protection for the remnants of the Beltrán Leyva Organisation, a once powerful cartel that is now in decline. The protection racket relies both on the territorial control the Zetas exert on the Atlantic coast of Mexico and their superb military abilities. The Beltrán Leyva Organisation's zone of influence is on the South West of Mexico, but they have operations in Quintana Roo, on the South East. While these latter operations might affect the Zetas' control over that area and, to some extent, minimise the possible revenue from the drug business, the connection between the two organisations is profitable because they partnered up to fight the Sinaloa Cartel, the largest and more dominant criminal organisation in Mexico at the moment.

The relatively loose connection between extortion dynamics and territory in Latin America is perhaps better exemplified by the pervasiveness of extortive kidnapping. Latin America accounts for 66 % of the extortive kidnappings worldwide (OEA, 2012). The planning and execution of extortive kidnapping do not depend much on the control of the territory, but on infrastructure and resources, especially when the victim is a high-value target. Territory, however, could become a strategic asset. In 2013, for example, the FARC kidnapped 16 people working for companies in the extraction sector, in spite of the tight control military forces keep over the extraction activities (Fundación Paz y Reconciliación, 2013). These kidnappings occurred both inside and outside their zone of influence. For the execution of the kidnapping, territorial control was subordinate to other aspects such as intelligence on the target and quickness in carrying out the abduction. Territorial control, however, is an important resource for the FARC, once the target has been kidnapped, for the victim is brought into their zone of influence and is usually moved through different locations, depending on the likelihood of military rescue operations (Fundación Paz y Reconciliación, 2013; Leech, 2009).

Extortive kidnapping in the region is practiced both by large and small criminal organisations. In Colombia, between 20 and 30 % of the kidnappings are carried out by small criminal organisations, with no affiliation to guerrillas or paramilitary movements. Because they usually lack the infrastructure and resources of large organisations, these small ERSs opt for more cost-effective modalities. One that is gaining increasing popularity across the region is called "express kidnapping" (Oropeza, 2015). The victim is abducted for a few hours, e.g. when boarding a cab, and taken to their place of residence or business, so the perpetrators can steal whatever they have, or to an ATM, where victims are forced to withdraw money from their accounts. Sometimes a small ransom that can be easily paid is demanded from companies or family of the victim.

Extortion is also particular in the region because criminal organisations usually do not have relatively homogeneous control of the territory. In Rio de Janeiro, for example, Comando Vermelho's control of the territory is limited by the urban layout of the favelas. Favelas are scattered across the city, but have clearly delimited borders with contiguous urban areas. This has created an interesting crime landscape in Rio. Within these borders, criminal gangs have successfully developed protection rackets that range from basic tax-like extortion to the monopolistic provision of social services, such as van transportation or cable TV, which the gang controls (Masciola, 2015). Beyond these borders, criminals lack the power to exert territorial control and heavily resort to crimes of opportunity. The strong link between these urban gangs and the favelas has made the pacification campaigns carried out by the Brazilian Government work almost like a shell game. Instead of a steadily expanding the pacification front, armed forces pick geopolitically important favelas, such as those around the city centre, and force the criminal gangs to relocate, usually to favelas in the southern outskirts of Rio, where they can still continue their criminal activities (Felbab-Brown, 2011).

Guerrillas across the region experience a similar situation. These criminal organisations usually control rural areas of the countries in which they operate. The success of this territorial control depends on geographical features associated with mobilisation, accessibility and camouflage. In Colombia, for example, guerrilla movements have, in part, been successful due to the particular features of the Andean mountain range. The Andes splits into three in the South of Colombia and cuts across the whole country. The rural areas of the FARC control are mostly around these three mountain ranges because of strategic advantages provided by the landscape. Apart from topographic conditions, these areas are mostly tropical forests, which makes access to them even more difficult. Protection rackets usually cover rural population and urban settlements located within these areas. The particular geographic conditions allow these organisations to perform extortion and other types of crime, minimising the risk of confrontation with governmental forces.

The uneven control of the territory has also made the ERSs in the region to specially target companies that are forced to use this territory. Extraction and transport industries are common targets. The Zetas, for example, have a combined scheme of petroleum rusting and extortion. Reports of PEMEX, the Mexican petroleum company, reveal that, between 2007 and 2012, the company suffered 1267 thefts, at a cost of $427 million. Large part of this oil was presumably stolen by the Zetas. In turn, this criminal organisation extorts companies in the North East of Mexico with 10 % of the contacts for gas extraction (Grayson & Logan, 2012). The FARC in Colombia also target extraction companies. In 2013, this guerrilla movement dynamited 108 energy towers and 259 pipelines, as a way to pressure these companies to pay extortion. They have also targeted machinery, vehicles and facilities in 44 occasions (Fundación Paz y Reconciliación, 2013). The maras have also developed a widespread extortion racket in the transport sector in the Northern Triangle. In Guatemala, for example, around 200 bus drivers were killed between 2005 and 2011, as a means to pressure transport companies to pay extortion. These gangs charge about $25 per bus each week. Each line has around 200 buses, which means that the gangs are pocketing around $500 for each line every week (Dudley, 2011). In Honduras, the maras earn around $16 million a year for extortion of the transportation sector (Cawley, 2013).

Criminal organisations have also specialised in making a transition in extortion dynamics from geographical to social spaces, such as local markets. In Latin America, face-to-face transactions still dominate the economy and social spaces such as local markets constitute a perfect target for extortion. They are very large and have significant amounts of transactions, usually in cash, occurring daily. Extorting sellers does not require significant resources or infrastructure and is difficult to identify. Chances of reporting are also very low. The maras in Tegucigalpa charge around $15 weekly to every shop in local markets. There are around 14,000 shops in 16 markets, which means that these criminal organisations are pocketing around $10 million from these extortion rackets (Cawley, 2013).

There is a large diversity of criminal organisations in Latin America carrying out extortion activities. These organisations are all similar in that they emerge and operate in a widespread context of exclusion, deprivation and violence, typical of the region. Unlike traditional mafia-type organisations, ERSs in Latin America do not depend entirely on territorial control for extortion. These organisations benefit from a pervasive context of criminality in which different sources of income can be found. ERSs in the region have developed a complex hierarchical structure of competition and collaboration that is mostly determined by the participation in the drug business, which is, by far, the most profitable criminal activity in the region (UNODC, 2011). Different types of extortion are articulated and carried out depending on the ERS's position in this criminal power structure.

2.5 Germany

According to the most recent report on organised criminality provided by the German Bundeskriminalamt (2013), there are two main groups involved in the practice of extortion rackets in Germany: outlaw motorcycle gangs and mafia-type organisations, especially the Italian and the Russian. Transnational outlaw motorcycle gangs have become a major problem, in terms of organised criminality. In several countries, these gangs have proven participation in criminal activities such as drug and gun trafficking (Barker & Human, 2009), but have always hidden behind the motorcycle club façade. These clubs are divided into chapters that are independent, but all chapters depend on the mother chapter for any major organisational decision. This alleged independence created significant difficulties in the prosecution of the entire gang, for it is often argued that crime is deviant not sanctioned action of the prosecuted chapter. The time taken for affected countries to update the normative framework in order to counter this type of criminal organisation has been used by most outlaw motorcycle gangs to articulate large well-connected criminal networks around the world (Barker, 2011).

In Germany, there is recorded presence of four major outlaw motorcycle gangs: Hells Angels, Bandidos, Gremium and Mongols (German Bundeskriminalamt,

2013); these groups were responsible for 11 reported cases of extortion (Bundeskriminalamt, 2013, S. 18). They, however, do not seem to have a major impact on criminality statistics at the moment. Their overall activities, including other kinds of crimes, in which participation is higher, make up for less than 10 % of organised crime in Germany.

Italian mafia has a documented presence in Germany. Several members of the Cosa Nostra, 'Ndrangheta and Camorra have been killed in Germany—e.g. in Duisburg in 2007 (Piller, 2007; Schilder, 2007) or arrested—e.g. near Lake Constance in 2015, where eight 'Ndrangheta suspects were detained, as they had been prosecuted by the Anti-mafia Directorate of Reggio Calabria for years (Braun, 2015). Yet, diasporas of Italian mafias do not seem to have engaged in extortion practices within the German territory. These groups were responsible for 11 reported cases of criminality, but most of them were not associated with extortion (Bundeskriminalamt, 2013, S. 18). Russian-language organised crime groups, on the other hand, were responsible for 30 reported cases, which, even though is almost three times the amount of Italian organisations, makes up for less than 5 % of organised crime reports in Germany (Bundeskriminalamt, 2013, S. 19). As "extortion racket" is not a reporting category of its own in German criminal statistics, it could be estimated that less than a half of the 30 reported cases were linked to extortion. The latter is a subcategory of "crime against property" and of "crime in economic life", which, together, add to only 14 cases in the main activity fields of organised crime groups from the countries of the former Soviet Union.

Unlike the situation in Italy, extortion rackets in Germany, due to their rarity, do not lend itself to statistical analysis. The numbers recorded in earlier reports of the German Bundeskriminalamt are approximately the same as the ones reported above for 2013. This stands in stark contrast to traditional zones of influence of these criminal organisations in the Italian *mezzogiorno*, where there were more than 100 reported cases of extortion per year between 2006 and 2013.[3]

2.6 Summary

This chapter has provided some evidence for the diversity of extortion practices and ERSs around the world. While extortion is often associated with traditional and popular mafia-type organisations, extortion carried out by the Italian, Japanese and Russian mafias can diverge significantly on crucial factors, for example, the social and political legitimation of the extortion. In addition to these traditional criminal organisations, several other relatively newer criminal groups with local and international reach are also engaged in different forms of extortion. These groups also display distinctive features, both in their organisation and in their relationship with

[3] Source: http://dati.istat.it/Index.aspx?DataSetCode=DCCV_DELITTIPS#, downloaded September 5, 2015, 09:48. The data from this source are not exactly comparable to the German data above, as the numbers for Italy refer to "tipo di delitto: Associazione di tipo Mafioso".

victims. Motorcycle gangs, for example, are not entirely criminal, while most ERSs in Latin America do not require exclusive territorial control for extortion. The next chapter tries to make sense of many of these differences through a typology that addresses the main contextual and organisational features of systemic extortion.

References

Alexeev, M. (1995). *The Russian underground economy in transition*. Report to the National Council for Soviet and East European Research, Title VIII Program. Retrieved from https://www.ucis.pitt.edu/nceeer/1995-809-04-Alexeev.pdf

Barker, T. (2011). American based biker gangs: International organized crime. *American Journal of Criminal Justice, 36*, 207–215.

Barker, T., & Human, K. (2009). Crimes of the big four motorcycle gangs. *Journal of Criminal Justice, 37*(2), 174–179.

Block, A. (1980). *East side, west side*. New Brunswick: Transaction Publishers.

Braun, J. (2015). Schwerer Schlag gegen die Mafia. *Südkurier, 154*, 10.

Bundeskriminalamt. (2013). *Organisierte Kriminalität. Bundeslagebild 2013*. Wiesbaden.

Burton, F., & Burges, D. (2007). *Russian organized crime*. Stratford. Retrieved from https://www.stratfor.com/weekly/russian_organized_crime

Cawley, M. (2013). La Extorsión Obliga a Cerrar más de 17.000 Negocios en Honduras. *InSightCrime*. Retrieved from http://es.insightcrime.org/noticias-del-dia/la-extorsion-obliga-a-cerrar-a-mas-de-17000-negocios-en-honduras

Chalk, P. (2011). *Latin American drug trade*. Santa Monica: RAND Corporation.

Dean, M. (2008). Sold in Japan: Human trafficking for sexual exploitation. *Japanese Studies, 28*(2), 165–178.

Dudley, S. (2011). InSide: The most dangerous job in the world. *In SightCrime*. Retrieved from http://www.insightcrime.org/investigations/inside-the-most-dangerous-job-in-the-world?highlight=WyJtYXJhIiwiJ21hcmEiLCInbWFyYSciLCJtYXJhJ3MiLCJidXNlcyJd

ECLAC. (2014). *Social panorama of Latin America*. Santiago de Chile: ECLAC.

El Espectador. (2014). Guerrilleros de las Farc entregaban "paz y salvo" a víctimas de sus vacunas. *El Espectador*. Retrieved July 29, 2014, from http://www.elespectador.com/noticias/judicial/guerrilleros-de-farc-entregaban-paz-y-salvo-victimas-de-articulo-509933

Felbab-Brown, V. (2011). *Bringing the State to the Slum*. BROOKINGS. Retrieved from http://www.brookings.edu/~/media/research/files/papers/2011/12/05latin america slums felbab-brown/1205_latin_america_slums_felbabbrown.pdf

Fundación Paz y Reconciliación. (2013). Cómo es Eso de Negociar en Medio del Conflicto. Retrieved from http://www.pares.com.co/wp-content/uploads/2013/12/Informe-Farc-2013.pdf

Gans-Morse, J. (2012). Threats to property rights in Russia: From private coercion to state aggression. *Post-Soviet Affairs, 28*(3), 263–295.

Grayson, G., & Logan, S. (2012). *The executioner's men*. New Brunswick: Transaction Publishers.

Hignett, K. (2012). We had to become criminals, to survive under communism: Testimonies of petty criminality and everyday morality in late socialist Central Europe. In M. Ilic & D. Leinarte (Eds.) *The Soviet past in the post-socialist present. Methodology and ethics in Russian, Baltic and Central European oral history and memory studies*. New York: Routledge.

Hill, P. (2006). *The Japanese mafia*. Oxford: Oxford University Press.

Hill, P. (2014). The Japanese Yakuza. In L. Paoli (Ed.), *The Oxford handbook of organized crime*. Oxford: Oxford University Press.

Hongo, J. (2008). Law bends over backward to allow "Fuzoku." *The Japan Times*. Retrieved from http://www.japantimes.co.jp/news/2008/05/27/reference/law-bends-over-backward-to-allow-fuzoku/#.VXm0LRNVhBc

Kaplan, D., & Dubro, A. (2003). *Yakuza*. Berkeley: University of California Press.
Leech, G. (2009). *Beyond Bogotá*. Boston: Beacon.
Levada Analytical Center. (2013). Russian public opinion 2012–2013. Retrieved from http://www.levada.ru/sites/default/files/2012_eng.pdf
Levitsky, S., & Way, L. (2010). *Competitive authoritarianism*. Cambridge: Cambridge University Press.
Masciola, S. (2015). Matthew B. Ridgway Center for International Security Studies, South America: Brazil. Retrieved from http://research.ridgway.pitt.edu/topics/gang-intro/south-americabrazil/
McCurry, J. (2008). Japan's lost decade. *The Guardian*. Retrieved from http://www.theguardian.com/business/2008/sep/30/japan.japan
McDermott, J. (2014). The FARC and the drug trade: Siamese twins? *In SightCrime*. Retrieved from http://www.insightcrime.org/investigations/farc-and-drug-trade-siamese-twins
OEA. (2012). *Report on citizen security in the Americas 2012*. Washington, DC: OEA.
Oropeza, V. (2015). Secuestro Express y Rescate en Divisas, un Delito en Auge en Venezuela. *El Nuevo Herald*. Retrieved from http://www.elnuevoherald.com/noticias/mundo/america-latina/venezuela-es/article25390987.html
Otis, J. (2014). *The FARC and Colombia's illegal drug trade*. Wilson Center. Retrieved from http://www.wilsoncenter.org/sites/default/files/Otis_FARCDrugTrade2014.pdf
Paoli, L. (2003). *Mafia brotherhoods*. Oxford: Oxford University Press.
Piller, T. (2007). Von Kalabrien in die Welt. *Frankfurter Allgemeine Zeitung, 189*, 7.
Rankin, A. (2012). 21st-century Yakuza: Recent trends in organized crime in Japan—Part 1. *The Asia-Pacific Journal: Japan Focus*, *10*(7). Retrieved from http://www.japanfocus.org/-Andrew-Rankin/3688/article.html
Schilder, P. (2007). Mit den Mitteln der Mafia. *Frankfurter Allgemeine Zeitung, 189*, 7.
Schneider, F., & Enste, D. (2002). Hiding in the shadows. *Economic Issues, 30*. Retrieved from http://www.imf.org/external/pubs/ft/issues/issues30/
Taylor, B. (2011). *State building in Putin's Russia: Policing and coercion after communism*. Cambridge: Cambridge University Press.
The Economist. (2008). Lenders of first resort. *The Economist*. Retrieved from http://www.economist.com/node/11413090
TRANSCRIME. (2008). *Study on extortion racketeering the need for an instrument to combat activities of organised crime*. Retrieved from http://ec.europa.eu/dgs/home-affairs/doc_centre/crime/docs/study_on_extortion_racketeering_en.pdf
UNODC. (2011). *Estimating illicit financial flows resulting from drug trafficking and other transnational organized crimes*. Viena: UNODC.
UNODC. (2013). *Global study on homicide 2013*. UNODC. Retrieved from https://www.unodc.org/documents/gsh/pdfs/2014_GLOBAL_HOMICIDE_BOOK_web.pdf
UNODCCP. (2000). *World Drug Report 2000*. UNODCCP. Retrieved from http://www.unodc.org/pdf/world_drug_report_2000/report_2001-01-22_1.pdf
Volkov, V. (2002). *Violent entrepreneurs: The use of force in the making of Russian capitalism*. New York: Cornell University.
Volkov, V. (2014). The Russian mafia: Rise and extinction. In L. Paoli (Ed.), *The Oxford handbook of organized crime*. Oxford: Oxford University Press.

Chapter 3
Basic Dynamics of Extortion Racketeering

David Anzola

3.1 Introduction

The emergence and prevalence of systemic extortion depends on particular institutional arrangements, produced as adaptive responses of the interaction between different social actors. The following typology characterises these arrangements in six categories, involving three main social actors: criminal organisations,[1] state and civil society. While the six categories focus on institutional arrangements that foster systemic extortion, the typology could be loosely described as presenting two categories focusing on each of the three main social actors: state, criminal organisations and civil society (see Fig. 3.1). The predominant social actor in each category derives from the relative importance of that actor for the specific aspect of the institutional arrangements addressed by each category.

The first two categories of the typology focus on whether the state is able to meet the citizen's demands for security. The first category, "state responsiveness", enquires about the institutional capacity for response of the state. It discusses the state's ability and will to exercise control over the everyday activities of citizens. The second category, "dissociation from the state", addresses the implications of particular instances where there is a willing disassociation of certain social groups from governmental institutions. In the third and fourth categories, attention is shifted towards the crimi-

[1] The notion "organised crime" is sometimes contested due to its generality and the fact that sometimes it theoretically attributes to the criminal organisation a level of cohesion and stability that is rarely seen in practice (Paoli & Vander Berken, 2014). In this case, because the focus is on systemic extortion, the notion "organised" is used referring to criminal structures that are relatively widespread, remain reasonably stable over long periods of time and, most importantly, are perceived by other social actors, who act according to those perceptions. Because the focus is on institutional arrangements generated from adaptive responses of different agents, this last factor is fundamental for the analysis.

D. Anzola (✉)
Department of Sociology, Centre for Research in Social Simulation,
University of Surrey, Guildford, Surrey GU2 7XH, UK
e-mail: d.anzola@surrey.ac.uk

C. Elsenbroich et al. (eds.), *Social Dimensions of Organised Crime*,
Computational Social Sciences, DOI 10.1007/978-3-319-45169-5_3

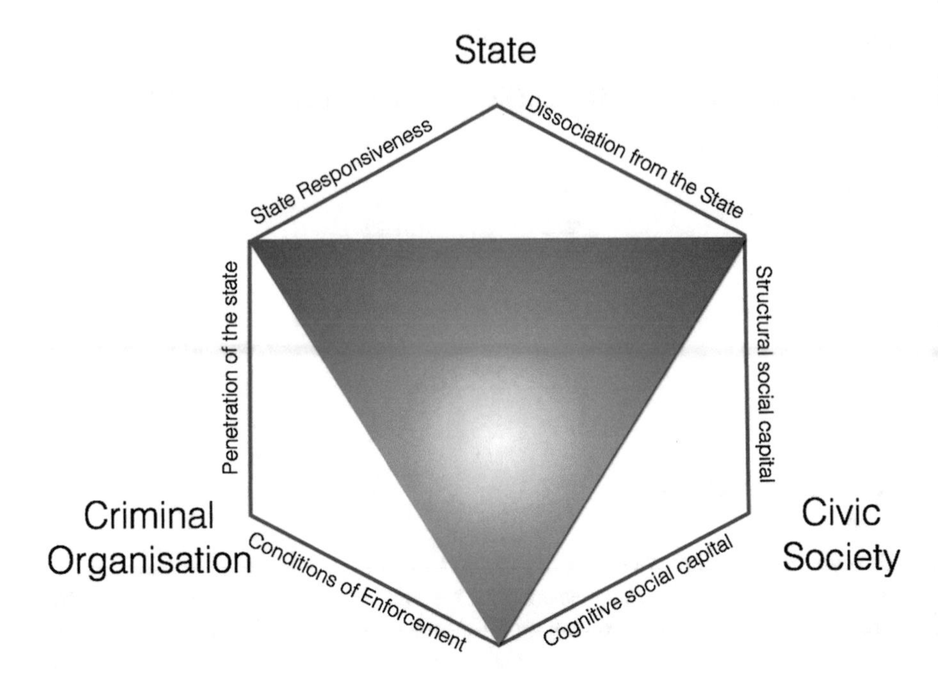

Fig. 3.1 The dimensions of a typology of ERS. The three social actors are the state, civil society and criminal organisation. For each actor two subcategories are developed through which their role in an ERS and their interaction with other actors are categorised

nal organisation and the conditions that enable its acquisition of power. The third category, "institutional penetration", analyses the way criminal organisations can permeate traditional institutions and practices. The fourth category, "conditions of enforcement", enquires about the factors allowing criminal organisations to achieve and maintain its institutional penetration. Finally, the last two categories address the victims, linking social connectedness and vulnerability. Social connectedness is approached using the concept of social capital. The fifth category, "cognitive social capital", focuses on emotional and cognitive processes that could foster systemic extortion by reducing the victims' sense of self-control and their access to social support. The last category centres on structural determinants that shape social resources to which a victim has access. It enquires about the objective external conditions of the social connectedness. It is labelled, accordingly, "structural social capital".

3.2 The State

3.2.1 State Responsiveness

Systemic extortion dynamics often depend on the exploitation of strategic spaces, both geographical and social, by the criminal organisation. The character of this exploitation, however, is sometimes muddled by a misleading interpretation of the spatial dimension of criminality. Extortion, and crime in general, has frequently been conceptualised suggesting that there is a power void or gap within a circumscribed space which is filled by the criminal organisation (e.g. Sung, 2004). While a weak normative framework and the lack of governmental institutions can certainly foster the emergence of criminality, referring to a power void creates confusion, for it implies that "presence" is the most important variable. Yet, in contexts of systemic extortion, where the activities of the criminal organisation are usually linked to the everyday activities of citizens in the overworld, state agents need not be absent and, sometimes, are actually crucial for the emergence and consolidation of the criminal organisation (see Pagden, 1990 for an example).

This misrepresentation of the role of governmental agents in systemic extortion is, in part, due to the fact that the notions "state presence" and "power void" presuppose a relatively static conceptualisation of the spatial dimension. Several concepts associated with space are taken as subordinate or by-products of general forms of interaction and social organisation. The concept "territory" in modern political thinking, for example, is often a subordinate of political and cultural aspects associated with the concept of nation-state. Its role is almost circumstantial. These static conceptualisations of spatial dimension lead to thinking of space as something that can be objectively partitioned, providing inadequate foundations for concepts such as "monopoly of violence" and "legitimacy" (Elden, 2013). That is particularly problematic when analysing extortion. The legitimacy of a political system, for example, is often associated with an alleged uncontested monopoly of violence. This, however, would be the case only if major institutional actors can interact with each other exclusively through competition. Yet, in different forms of criminality, including extortion, the coexistence of several actors, either legal or illegal, with the ability to exercise widespread violence is common. In Japan, for example, the Yakuza has historically worked alongside the Japanese Government and right-wing elites. More recently, it has also established working relationships with the Chinese triads, which operate, for example, in some areas of the entertainment sector in Tokyo, traditionally controlled by the Yakuza (Hill, 2014).

In extortion dynamics, then, the spatial dimension is important not much in terms of presence, but of responsiveness. Space can influence governmental responsiveness insofar as it is a product of social interaction (Gieryn, 2000; Thrift, 1996). Under such approach to space, the absence or inadequate response of governmental agents is but one factor shaping and being shaped by everydayness. In the case of systemic extortion, it is important to analyse how different power dynamics between victims, perpetrators and other individual and institutional actors allow for the emergence and maintenance of extortive dynamics, associated with certain social and geographical spatial configurations. The nature and character of power are linked to particular contexts where power asymmetries are identified and reacted upon by the participating actors (Foucault, 1982), either by physical displays of contentious politics or social control (Solt, 2012) or through a symbolic framework, operating through subtler mechanisms, e.g. media (van Dijk, 2012).

The rural–urban divide is a space-based power asymmetry that could prove difficult to eradicate, even for a responsible, committed and willing state. Urban and rural areas are often differentiated by demographic factors, such as population density. Yet, especially in underdeveloped countries, the difference is mostly characterised by significantly lower access to political, social and economic resources in rural populations (The World Bank, 2014). This limitation in resource access has been widely exploited by criminal organisations such as guerrilla movements. The FARC, for example, has established several protection rackets in rural areas of Colombia (Ferro & Uribe, 2002). In some areas where they have full control, the organisation works almost as a "family" movement. Families of FARC members usually provide intelligence or goods, such as food. Increasing state presence is unlikely to undermine the power of the criminal organisation, for it does not target its source, which is the existence of kindship relationships within that space.

Spatial power asymmetries need not be entirely linked to geographical features. Markets, for example, can equally defy governmental control because of the pace, volume and diversity of interactions. Different types of markets pose different challenges for control. The global financial market fosters different criminal activities because (a) criminal activities are usually carried out by people inside the market during their everyday job (Marroco, Fisher, & Kishida, 2008), (b) full information about the transactions is rarely available and (c) institutional responses against white-collar crime tend to be rather weak (Nelken, 2012). Financial agents providing services to criminal organisations can be easily extorted into continuing working in illegal transactions. In cases where there is no immediate awareness of the source of the assets, financial agents might have problems demonstrating their innocence and, even if they do, their reputation could be irreparably damaged (Friedrichs, 2010; Levi, 2014).

Extortion dynamics in more basic markets, such as urban agricultural markets, are somewhat different. The *Mbare Musika* market is the most important agricultural market in Harare, Zimbabwe. Around 2500 trader and tens of thousands of visitors converge in the market daily. The value of goods traded ascends to over 36 million dollars annually (Muzulu, 2014). Most transactions in the market are carried out using informal methods. Criminal organisations, e.g. the Chipangano gang, have developed extensive extortion rackets that take advantage of the availability and accessibility to a large amount of informal transactions in a relatively small physical space.

Using space as a production resource when analysing extortion contextualises the phenomenon of interest; it sets physical and social boundaries. The acknowledgement of these boundaries allows identifying constraints on the use of social and individual resources, based on roles, beliefs and expectations of the participants. Spatial boundaries determine social phenomena by linking expectations with actions in the production and reproduction of power structures in particular contexts. Extortion in financial markets, as mentioned, depends on individuals that are highly positioned in the power structure because of their knowledge and skills. This position provides, among others, advantages when facing law enforcement institutions.

Under this perspective, power voids are not really voids inasmuch as they usually reflect particular institutional arrangements, mediating tensions between the objective power structure and the expectation of participating social actors. In Rio de Janeiro, for example, pacification campaigns carried out by the Brazilian government in the *favelas* have not followed a uniform pacification front, but have instead focused on geo-strategically located *favelas*, because of the more general political and economic implications of keeping these *favelas* away from the reach of organised crime (Felbab-Brown, 2011). The construction of expectations is also symbolic. It is noticeable, for example, how the recent wave of immigrants to the European Union has been reported using the notion of "migrants" instead of "refugees" by most media outlets (Malone, 2015), because of the implication this might have, among other things, on the political and legal responsibilities of the European countries.

3.2.2 Dissociation from the State

Systemic extortion may depend, it was argued, on the ERS's ability to exploit strategic social or geographical spaces. This exploitation was associated with the development of power structures that allow the criminal organisation to impose its will on the victims. The first category focused on power relationships within a unified social or geographical space. Systemic extortion, however, could also occur in situations where there are processes of dissociation of the victim from the state, regardless of the power, presence and responsiveness of governmental institutions. The power dynamics that allow for extortion on this second category come from the fragmentation of the power structure, and therefore of social and geographical spaces, due to willing or conscious actions of the victims. While in some cases poor state response and victim's dissociation can occur simultaneously, it is important to keep them separate, for, as it will be shown below, there are populations with a high degree of vulnerability under the category of dissociation from the state that would barely be covered by the category of state responsiveness.

Probably the most straightforward case of increased vulnerability due to the position of the victim in the power structure are members and associates of criminal organisations. Extortion is a widespread mechanism of interaction between and within criminal organisations. On the one hand, it often guarantees the permanence of less committed members or associates; on the other, it prevents disruption of business in lower scale criminal organisations. In Colombia, for example, different actors in the drug business operate under a widespread extortion racket established by guerrilla movements, particularly the FARC, given the control this guerrilla movement has of the areas where coca is grown (McDermott, 2014).

While it might not be in the interest of the state to protect the criminals, there are vulnerable populations that are caught up in this institutional power conflict. In not fully criminally oriented organisations, e.g. motorcycle gangs, members might get involved in criminal activities against their will, without a safe way out. In turn, many civilians are extorted of forced to work for criminal organisations. Most coca growing, especially in Peru and Bolivia, for example, is done by local residents, which are either forced to grow coca or just do it because it is the only mean of subsistence they can find (Briceno & Bajak, 2015).

Vulnerable populations that eventually dissociate from the state need not be at the bottom of the power structure. There is, for instance, a particular form of corporate extortion in Japan, usually known as *sōkaiya*. In this type of extortion, criminals extort money from well-known companies or their managers, with threats as simple as disrupting shareholder meeting. In the case of *sōkaiya*, Japanese companies put up with the extortion and dissociate themselves from the state in order to keep a favourable institutional setting. Institutional means of control and accountability in corporate Japan are very weak. Even with the losses generated by extortion payments, companies can still thrive economically in a way that is not possible in other more regulated institutional settings.

Apart from the position in the power structure and the level of involvement in criminal activities, vulnerability might also be linked to social representations, cultural factors and institutional conditions of neglect. Systemic extortion is common in the prostitution business and it takes many forms, e.g. street-level pimping and human and drug trafficking (Human Rights Watch, 2000). In spite of high levels of victimisation, prostitutes are forced to isolate themselves mostly due to the negative moral representation of their profession (Bruckert & Hannem, 2013). Stigmas associated with prostitution derive in institutional neglect and repeat victimisation, both by state agents and general society (Farley & Barkan, 1998). Some vulnerable populations can be stigmatised in more than one way. Ethnic minorities with criminal records, for example, are frequently stigmatised both because of their ethnicity and their legal situation (Campbell & Deacon, 2006). Stigmatised population are a common target of extortion because of the vulnerability derived from the stigma, which in most cases is not necessarily linked to an association with criminal organisations.

Ethnicity is a cultural factor that can often lead to vulnerability. Successful integration of migrant communities, for example, depends on processes of acculturation. In some occasions there are technical, i.e. language, and cultural, e.g. religion, factors disrupting the process, which results in low levels of integration, civil participation and general well-being (Ho, 2014; Schnittker, 2002). These migrant groups might dissociate from the state because they are dependent on someone else for interaction with the outside world or because they are willingly rejecting the new culture (Phinney, 1990). Not fully acculturated communities are regular victims of extortion because of their level of isolation or autonomy, but also because ethnicity sometimes provides means of legitimation and control for the criminal organisation. Some victims tend to legitimise extortion by not taking it as such, but as a practice that fits their custom or traditions (Chin, 2000).

Dissociation can originate as a response to long-standing processes of institutional neglect. Successful levels of integration depend on the way political institutions deal formally and informally with processes of inclusion and exclusion. A democratic system might be designed in such a way as to formally or informally exclude a portion of the population (Holden, 2006). Voting is a typical example of the former. The latter is usually produced by the inappropriate functioning of the institutional setting. Income inequality, for example, is linked to a decrease in political interest, discussion and participation (Putnam, Leonardi, & Nonetti, 1993). In the long run, it might generate widespread feelings of apathy, which often translate in dissociation from political institutions (Solt, 2008). In places like Latin America, where exclusion and inequality are chronic, disadvantaged groups' dissociation from the state combines lack of interest in political participation, lack of trust in governmental institutions and a fairly inefficient law enforcement governmental apparatus. As a result, different types of victimisation, including extortion, are rampant, while rates of denouncing, prosecution and conviction are low.

This second category addressed types of vulnerability that emerge from the fragmentation of social and geographical spaces. While poor or inadequate state response has more profound effects on those in lower or peripheral places in the power structure, willing dissociation can create fragmentation in any part of the power structure. Japanese corporations, for example, have significant economic and political capital. The extent and implications of dissociation are context dependent. Migrant communities, for example, are likely to achieve higher degree of integration when the host is a very multicultural society. Likewise, the degree of cultural separation affects both the chances of stigmatisation and dissociation. The contemporary narrative on terrorism, for example, depends on a particular geopolitical reconstruction of the East–West tension (Cowen & Gilbert, 2008). Identifying those fractures in the power structure and how they might develop into isolation or radicalisation are fundamental to understand adaptive responses that eventually allow for systemic extortion.

3.3 The Criminal Organisation

3.3.1 Institutional Penetration

Within a social or geographical space, the chances of systemic extortion depend on the level of institutional penetration[2] the criminal organisation can achieve. Institutional penetration can provide the ERS with reasonably stable formal or informal means of control over institutional resources and mechanisms that facilitate extortion. This control is provided, as mentioned, not by a power void, but by institutional arrangements involving both governmental institutions and civil society. If certain activities are not typified as extortion, for example, there is a formal

[2] Institutional penetration does not directly correlate with the level of activity of the criminal organisation. Two criminal organisations can differ in the level of institutional penetration while being similar in the type and extent of their extortion activities. However, it is likely that the organisation with lower level of institutional penetration is required to invest more resources into securing the temporal extension of the extortion. Identifying these institutional arrangements does not provide much insight into what type of extortion is possible, but into what type of institutional arrangements are likely to lead to tenable systemic extortion.

legal arrangement that benefits the activity of the criminal organisation. Likewise, if the victim finds normative reasons for complying with the criminal's demands, there is an informal individual arrangement that facilitates extortion. This section focuses on how ERSs can take over power structures due to particular institutional arrangements. The next category focuses on the particular enforcement of their will on these spaces.

Institutional penetration has to do with power and control; these can be acquired and exercised in many ways. The literature on corruption and organised crime is framed in such a way to make the state, to a certain extent, a passive victim of organised crime (Hellman, Jones, & Kaufmann, 2003; UNODC, 2010). This approach to systemic political corruption, however, has three problems: First, it neglects that power dynamics are often the result of bargaining processes between individual and institutional actors within or outside governmental institutions (Karstedt, 2014). In some cases, state agents play a direct role in extortion. In Latin America, for example, there are large and powerful criminal networks working from within the state, targeting both individuals and criminal organisations. In 2014, for instance, 16 police officers, including a lieutenant colonel, were arrested in Rio de Janeiro, accused of extorting small drug trafficking gangs (EFE, 2014). In 2015, another 46 were expelled from the force after being accused of extorting small shopkeepers in exchange for protection (EFE, 2015).

The view of the state as a victim of organised crime also leads to underestimate the state's capacity to counteract organised crime. After Putin's rise to power, for example, law enforcement institutions have pushed the Russian mafia away from its traditional zones of influence (Volkov, 2014). This change does not seem to be associated with a generalised reduction of criminality levels or an increment in trust in political institutions (Levada Analytical Center, 2013; Taylor, 2011), but with the transition of the Russian political system into a competitive authoritarianism (Levitsky & Way, 2010).

Finally, the view of victimisation of the state is also problematic for it does not recognise the fragmented character of both the criminal organisations and the governmental institutions. The concept of "organised crime", as mentioned, is problematic because it gives criminal organisations a sense of unity and structuration that does not exist in real life (Paoli & Vander Berken, 2014). The same can be said about the state. Functional, bureaucratic and spatial differentiation has an effect on the operation of governmental institutions. Interaction between criminal organisations and the state varies significantly, according to the level of decentralisation of democratic institutions, as well as the separation and relatively high level of specialisation of different governmental institutions. Criminal organisations have historically bribed and extorted individuals in democratically elected positions with votes (Gambetta, 1990; Varese, 2014). The accumulation of political capital, however, does not help criminal organisations, for example, when interacting with the justice system, since people holding positions in this governmental branch are usually not elected democratically. In turn, because of the presupposed legitimacy in the exercise of violence, law enforcement agents, as attested by extortion rackets in Rio, can engage in forms of criminality that other governmental instances cannot.

The possible presence of powerful criminal networks within the state makes criminality a three-way relationship that blurs to a certain extent the distinction between public and private. The level of penetration of ERSs is determined by how the interests of the state, public and private criminal networks and organisations align over time and the type of relationship developed when there are common interests. In terms of reducing vulnerability, one aspect is whether different governmental institutions and factions are operating under relatively homogeneous goals. In some instances where these goals have been aligned, such as Russia and Georgia, governmental efforts to fight organised crime have shown noticeable results in a relatively short amount of time. In most cases, however, criminality is perpetuated through bribing and extortion of corrupt or criminal factions within the state.

Something similar happens with civil society. Traditional mafia-type organisations (MTO) furthered their role as brokers of social life by choosing partnership over competition in their relationships with actors in the civil society. The Japanese Yakuza, for example, has several so-called business brothers (*kigyō shatei*): individuals that are not officially connected to the organisation, but benefit from informal relationships, e.g. achieving cheaper production cost by avoiding the law (Hill, 2006). In occasions, these links have formal character. In the construction business in Italy, for example, the mafia had representation in the *comitati d'affari*, a group composed by politicians and entrepreneurs, which, for years, decided on the allocation of large-scale public works across the country (Paoli, 2003).

Criminal organisations can increase institutional penetration in civil society by diversifying the methods of extortion, without entirely moving away structurally from the services provided by the market or the government. Traditional MTO became successful brokers of social life by entering in markets in which there is an identifiable demand and competition, even if rigged, is possible. That helped making the distinction between their provision of good and services and extortion, to some extent, artificial for the consumers/victims. The concepts could be considered exclusive only by presupposing that the illegal provision by the criminal organisation is suboptimal, in comparison to a "perfect" provision by the state. Yet, that is hardly the case (Varese, 2014). Criminal organisations need not guarantee that they are the only one providing the service/good, nor that the service/good they provide is the best. They just need to make sure that victims/consumers perceived it as a plausible alternative. In some cases, given the level of inefficiency of the system and the victims' dissatisfaction and distrust in political institutions, the standard for "plausible" is very low.

The level of institutional penetration is, then, dependent on the possibility the criminal organisation has of achieving a power elite status (Wright Mills, 2000). Achieving this status hinges, initially, on the organisational capacity of the criminal organisation. Higher levels of specialisation allow for the control of larger and more complex institutional resources (Decker, Bynum, & Weisel, 1998). Deep and widespread control, however, is only possible if the criminal organisation incorporates or establishes links with individual or groups that broker access to key social resources the criminal organisation could not access itself, for example, the legal regulatory framework (Garay-Salamanca & Salcedo-Albarán, 2012). These brokers, placed in different positions of the power structure, could have a willing or an unwilling role in increasing the criminal organisation's institutional penetration, depending on, for example, whether interaction is based on bribery or extortion. They could also have a passive or active role, depending on, for example, whether the contact with the mafia is direct or via intermediaries.

3.3.2 *Conditions of Enforcement*

Once certain arrangements allow ERSs to penetrate specific institutional spaces, it is up to the criminal organisation to generate mechanisms through which its demands can be enforced. The mechanism chosen depends on the structural features of the criminal organisation, contextual social conditions and type of extortion. The possibility of long-lasting extortion depends on the resources criminal organisations and other actors involved have to invest for the exploitation of these institutional spaces, as well as the adaptive responses to institutional changes in the extortion landscape. Criminal organisations, for example, might encounter less institutional resistance when protection is genuine, but can equally change from a bogus to genuine extortion as an adaptive response to the emergence of other criminal organisations offering the service (Abadinsky, 1983).

Regardless of the type of extortion context, the existence of the criminal organisation is a social bad, leading to suboptimal results that need to be enforced by physical or symbolic means (Gambetta, 1990; Varese, 2014). The chances of paying are influenced by attitudes towards the criminal organisation and its perceived strength, both to impose to extortion and to punish the victim in case of non-compliance. A positive attitude towards the organisation and a perceived high strength increase the chance of paying; a negative attitude and a perceived low strength decrease it.

The perceived strength of the criminal organisation is a criterion that combines objective and subjective factors, both at the individual and the social level. When extortion is considered a matter of victimisation, judgements of the victim are linked to perceptions of vulnerability (Skogan & Maxfield, 1981). It is not so much about what the organisation does to enforce its will, but about the fear extortion generates in the victims. Fear is the result of judgements about exposure to risk, anticipation of serious consequences and lack of effective means of control over crime, e.g. chances of being able to escape (Killias, 1990). These three dimensions are evaluated by victims using both cognitive and emotional elements. The former are associated with evaluations of likelihood and cost, and the latter, with emotional responses based on mood or the vividness with which the occurrence of extortion can be recreated (Loewenstein, Weber, Hsee, & Welch, 2001). Emotional elements have been shown to greatly distort judgements of vulnerability (Jackson, 2008), making it possible for an organisation with objectively limited means to enforce the extortion.

Attitudes towards the criminal organisation depend, among other things, on the moral foundations for concepts such as fairness and righteousness and their applications to particular episodes of victimisation, on the reconstruction of the criminal's otherness and the physical and symbolic contexts associated with victimisation episodes. Regarding the first one, not everyone considers the same action amounts to victimisation or crime. That is particularly true in situations where violence is mostly symbolic, e.g. gender violence (Fischbach & Herbert, 1997). The criminal's otherness is important in fostering or hindering negative judgements. The acknowledgment of difference is one of the reasons why, for example, there are widespread prejudices linking violence with migration (Tileaga, 2006). Finally, context is important for it points to the individual's beliefs about a fully functional society. Signs of urban decay, e.g. broken windows and graffiti, are usually mentally associated with disorder and crime, event though that need not be the case (Villarreal & Silva, 2006).

While individual attitudes and perceptions about criminal organisations can significantly vary within a group of victims, there are some general aspects of the extortion relationship that can have widespread effects on these attitudes and perceptions. The conditions for enforcement are initially linked to the type of institutional penetration. The broker role adopted by traditional MTO gave them a central position, connecting diverse social institutions and individuals. This allowed them to develop strong and long-standing bonds and partnerships with people in government and civil society, which facilitated the exploitation of institutional spaces due to the lack of resistance. The apparent ubiquitous and sometimes positive character of the MTO participation in social life is, in part, one reason why the anti-mafia measures have developed an ethical component aimed towards local population (La Spina, 2008).

Traditional MTO have also positively affected attitudes and perception of strength through a strategic use of violence. These organisations are not reluctant to use violence to enforce their demands. Yet, they have managed to keep a relatively low or good profile, due, in part, to particular efforts of moderation and self-regulation. Numbers for kidnapping in Italy, for example, have not steadily increased due to the pressure this type of crime puts on the delicate relationship criminal organisations have with each other and with the state (Dickie, 2008). Likewise, unofficial accounts suggest that, in Tokyo, the Yakuza has not been entirely pushed away from areas such as the entertainment sector, partly because the police considers that this criminal organisation helps maintaining low levels of petty crime (Hill, 2006).

The level of violence is partly associated with the nature of the service attached to the relationship of extortion. Attitudes towards the Italian mafia in the USA, for example, are more negative than in Italy, partly because these organisations did not take a role as brokers of social life. They, instead, got involved mostly in illegal and criminal activities (Varese, 2014). The low participation in the legal economy and the decisively abusive character of its extortion rackets has prevented the American mafia from developing the same normative nexus to civil society and governmental institutions of its Italian counterpart (Anderson, 1965). A factor that was also different for the American mafia is the ethnic background. Traditional MTO often appeal to ritualistic cultural factors. These factors generate cohesion and obedience within the organisation. They also strengthen their position within civil society, by reinforcing a social representation of shared practices, which facilitates recruitment and widespread loyalty, among other things. Shared ethnic background seems to be particularly important to consolidate the continuity of the criminal organisation over time.

Compliance with the criminal organisation's demands depends on a complex network of objective and subjective factors, related to features of the criminal organisation, the victim and the social context. The victim-perpetrator relationship does not have to be of antagonism. When it is, the effectiveness of the extortion depends on perceptions of vulnerability. Compliance is achieved by different combinations of symbolic and physical violence, which are linked to exercise of power either through processes of bargain and punishment (Lawler, 1992).

3.4 Civil Society

Extortion has some shared features with other forms of criminality, such as its dependence on the exercise of violence, while, at the same time, it displays some distinctive features, such as the development of trust networks that, to a certain extent, modify or regulate the exercise of violence. Moving from this general perception to a more throughout characterisation of extortion dynamics is difficult, for it is an extremely diverse phenomenon that can take many forms. It is usually understood as a social and economic transaction that is underlain by a coercive relationship between the parts. Yet, even in the most widespread case of protection rackets, the character of the transaction and the coercion mechanisms cannot be easily generalised.

Extortion is better understood as an umbrella concept, grouping a diverse set of practices that allows the perpetrator to take a position of enforcer in the social structure that, by its causes or consequences, is considered morally reproachable, economically inefficient and, in general, socially inconvenient. When it is systemic, the most important feature for an account of victimisation has to do with its extension in time. The typology uses the concept of social capital to capture the most relevant aspects of the temporal dimension of systemic extortion from two different angles: the way it affects, on one side, the cognitive or perceptual aspects of associational bonds or social activity and, on the other, the extent and intensity of these bonds. These two groups are labelled in the literature as "cognitive" and "structural" forms of social capital (Almedom, 2005; Harpham, Grant, & Thomas, 2002), which are the last two categories, respectively.[3]

[3] For a more thorough justification of the inclusion of the concept of social capital see Anzola & Elsenbroich (2015).

3.4.1 Cognitive Social Capital

The literature on social capital could allow explaining the prevalence of extortion by focusing on how systemic victimisation produces or takes advantage of low social connectedness. The threat of extortion is an instance of what in psychological and health literature is known as a “stressor”, i.e. a social demand for modification of the person’s behaviour (Carr & Umberson, 2013). In order to prevent negative effects on the victim’s well-being, different coping resources and strategies need to be implemented. Some of these resources rely on the person’s confidence in his or her ability to overcome the stressful situation (Pearlin, Menaghan, Lieberman, & Mullan, 1981). Systemic extortion, however, is problematic because it is a stressor that requires behavioural modifications to be sustained over long periods of time. The prevalence of extortion increases emotional distress, in the form of depression, anxiety and/or anger, leading to an increased perception of powerlessness and hopelessness (Ross & Mirowsky, 2006).

Along with coping resources associated with the person’s sense of control, stress can also be dealt with by the acquisition of social capital from which instrumental, emotional and informational assistance can be obtained. Systemic extortion, however, tends to target populations where access to the required support networks is hindered by objective and subjective conditions disrupting the formation and acquisition of social capital. Income inequality, for example, is an objective circumstance that hampers the access to appropriate coping resources and significantly affects overall mental health (Lynch & Kaplan, 1997). Income inequality can also affect coping resources indirectly, by negatively affecting the cognitive mediators that link emotional distress with the objective conditions of disadvantage generated by limited income. It can, for example, prevent a victim from coping with the distress, by affecting the victim’s beliefs on social justice (Wilkinson, 1992). In some cases, negative beliefs or feelings associated with objective conditions, such as income inequality, can be structurally amplified. The feeling of uneven social justice might be worsened by the sense of powerlessness generated by emotional distress, which might lead to even greater impacts on mental health and social capital (Harpham, Grant, & Rodriguez, 2004; Ross & Jang, 2000).

Sustained compliance in cases of systemic extortion is partly achieved due to the poor social capital most victims have before the extortive relationship is established. This lack of social capital, at a cognitive level, will likely be associated with impoverished mental health. The key to success, however, is likely linked to the fact that the threat of victimisation can reinforce the cognitive and emotional traits that increase the sense of powerlessness and hopelessness in the victim. Evidence for that is found in, perhaps, the most researched case of repeat victimisation: domestic violence.[4] Domestic violence is a type of victimisation with very low levels of

[4] As a point of conceptual contrast, domestic violence has the advantage of being a rather isolated form of systemic victimisation. It is clearly delimited within the boundaries of the household. In turn, the context and participants are likely to remain reasonably stable throughout the process.

report, in spite of the existence of relatively simple ways to avoid victimisation. The literature shows that the decision to terminate an abusive relationship is mediated by emotional and rational elements. Emotional aspects are particularly relevant when judging deviance and victimisation. Victims emotionally attached to their victimisers seem to be more willing to justify the victimisation or even be reluctant to acknowledge that the relationship is abusive (Felson, Messner, Hoskin, & Deane, 2002). Emotional responses to domestic violence that prevent reporting might also be reinforced by individual and social stereotypes surrounding this type of victimisation, e.g. victim-blaming (Yamawaki, Ochoa-Shipp, Pulsipher, Harlos, & Swindler, 2012). Rational elements seem to play a role when victims consider the presence of the victimiser has an important impact on their well-being, for example, when the victim depends economically on the victimiser or when reporting might expose involvement with illegal activities (Felson et al., 2002).

Rational and emotional judgments regarding the decision to terminate an abusive relationship could be significantly affected by the temporal extension of this type of victimisation. The literature acknowledges that one important factor preventing victims from escaping domestic violence is the mental health issues derived from episodes of repeat victimisation. The prevalence of domestic violence is linked to negative effects violence has on beliefs of self-worth and personal control, which, over time, lead to feelings of powerlessness and hopelessness (Fischbach & Herbert, 1997; Kennedy, Bybee, & Greeson, 2014). Adaptive responses to repeat episodes of victimisation end up in the loss of sense of personal control, which decreases the victim's ability to cope with the victimisation. The literature shows that, because of the loss of sense of personal control, informal support networks, such as family and friends, become fundamental at the moment of taking the decision to terminate an abusive relationship (Chang et al., 2010; Fugate, Landis, Riordan, Naureckas, & Engel, 2005). Victims of domestic violence that do not have this support are less likely to judge their situation as negative and seek help.

Victims of systemic extortion are likely to display similar inhibitors to those displayed by victims of domestic violence when deciding whether to report. The involvement in illegal activities is a straightforward concern, which is particularly relevant in cases of symbiotic extortion, where the victim and the perpetrator might be working in collusion. Victims might also be reluctant to report when the criminal organisation is involved in the provision of social and public services. Reporting the extortion in these cases might have a negative impact on the victim's well-being that might far outweigh the advantages. In cases of systemic extortion there might also be emotional inhibitors for reporting. As mentioned above, cultural and symbolic aspects could significantly affect the social representations of crime and deviance. Extortion carried out by Chinese gangs in New York is socially validated by the shared ethnic background. Likewise, the *oyabun-kobun* kinship institution of the Yakuza or the *omertà* code of the Sicilian mafia are linked to traditional and positively valued symbolic structures that, among other things, increase validation of the criminal organisation's activities (Ishino, 1953; Paoli, 2003).

While a reasonably good amount of literature can be found about the context of victimisation in systemic extortion, there is a lack of information linking the context to, first, prior deficiencies in the victim's mental health before the extortive relationship is developed, and, second, the boosting of these deficiencies or the development of additional ones due to the sustained stress and emotional distress generated by a constant threat of victimisation.

3.4.2 Structural Social Capital

The notion of structural social capital focuses on objective conditions of social connectedness. It enquires about the extent and intensity of this connectedness and the implications it has on people's well-being. Because of its macro character, structural social capital depends on social conditions of production and reproduction. It differs from other forms of capital, such as human and economic, in that, because it is a public and not a private good, it is created as a by-product of social interaction (Coleman, 1990). Due to this, the production of social capital cannot entirely rely on economic or political institutions. If social links are not there, it is likely that social capital will be underproduced.

Systemic extortion is particularly common in contexts where structural conditions hinder the development and accumulation of social capital. These conditions might come as a result of long social processes, where extortion is coupled with additional social, environmental and individual stressors, or be linked to specific life events, which produce immediate adaptive responses. Objective conditions of disadvantage could be worsened by the stress and emotional distress caused by sustained exposure, which, in time, might lead to social withdrawal (Kawachi, Kennedy, Lochner, & Prothrow-Stith, 1997). Important life events, e.g. violent victimisation, might also lead to social withdrawal. Individuals who have been victimised tend to have less social participation, partly linked to a decrease in trust in the community (Stafford, Chandola, & Marmot, 2007).

There are other important conditions of vulnerability beyond social withdrawal that can foster the emergence of systemic extortion. Some communities have been shown to have high levels of social cohesion in spite of relatively disadvantageous socioeconomic conditions (Fassaert et al., 2011; Villarreal & Silva, 2006). They, however, are vulnerable because there is not enough social connectedness outside the group. This is a factor captured by the distinction between bonding and bridging social capital. The former focuses on social connectedness between individuals that share the same background or display similar signs of belonging, and the latter on social connectedness at the institutional level or with individuals from different social groups (Putnam, 2000). A third type of social capital: linking social capital, could be added, to encompass social capital produced by social interaction through formal channels or official institutions (Szreter & Woolcock, 2004). The previous categories of the typology show that bonding social capital is not sufficient for preventing systemic extortion. Many social groups described in the section on dissociation from the state, for example, display high levels of bonding social capital. Yet, they are still easy targets for systemic extortion. Bonding social capital is fundamental for personal development and the sense of self. Yet, it only provides individuals with a relatively small range of material and symbolic resources. These resources are particularly limited when dealing with large institutional actors, including criminal organisations. It is unlikely, for example, that small communities have at their disposal communal resources to counter the institutional means for violence criminal organisations usually have.[5]

The addition of linking social capital seems to be particularly relevant for the context of extortion and victimisation, since different institutional channels have different degrees of legitimacy and social validation. Formal institutions are important because they usually provide access to more diverse and effective institutional resources. In the South of Italy, for example, trust in traditional political institutions as well as levels of civic participation and political engagement are low (CNEL/ISTAT, 2014). It has been suggested that this is a crucial reason for the emergence and persistence of MTO in the region (Putnam et al., 1993). The problem of low social capital in contexts of low civic participation and political engagement is that the generation of social capital cannot be met by traditional political institutions because of the disconnection between citizens and state.

In those instances where traditional political and economic institutions fail, other forms of institutional support are necessary. As mentioned, given the material and symbolic stronghold of criminal organisations in the South of Italy, anti-mafia policies tend to emphasise the importance of the moral component linked to generating awareness. Intermediary formal institutions of civic character, such as *addiopizzo*,[6] which are considered more trustworthy by the citizens (CNEL/ISTAT, 2014), seem to have successfully filled the gap between citizens and governmental institutions, by providing citizens with new institutional channels that are both powerful and trustworthy.

The existence of relatively centralised institutional resources is important to counteract the institutional power of the criminal organisations, which is significant, given the fact that several forms of extortion are not relegated to the underworld. Most ERSs are not just economic syndicates engaging in criminal activities with the goal of economic profit; they are power syndicates that use extortion as a mean of acquiring and maintaining power. Criminal organisations need to make strategic use of violence to avoid dissidence and appease social unrest, in the same way the state does. Because they might be competing with other institutional actors for power, avoiding dissidence and unrest takes a central position.

Criminal organisations in the extortion business end up relying on social control practices similar to those employed by authoritarian states. These practices seem to

[5] In a sense, however, the development of protection rackets could be understood as an increase in social capital, given the fact that institutional means of violence are developed by a private actor that might not be considered an outsider. Taking into account cases such as the Chinese migrants in New York, it is likely that the perception of this private monopoly of violence as originating from bonding or linking social capital has significant influence on judgements about crime and deviance. A monopoly of violence that emerges as bonding social capital is more likely to be positively valued by the victims.

[6] http://www.addiopizzo.org

regularly target different forms of civic association and tend to undermine social connectedness (Stenner, 2005). Authoritarianism undermines social capital by explicit displays of material and symbolic violence, both by those who hold power and those who independently help from the periphery. Criminal organisations might hurt a victim that refuses to pay protection or is likely to report extortion to the police. Successful control, however, depends on strategic infusions of distrust that undermine the creation of social capital from within civil society (Solt, 2012). ERSs use extortion to negatively affect the outcome of social exchange, in order to reinforce the belief on the need for their existence, while at the same time increasing distrust within civil society (Gambetta, 1990).

Criminal organisations do not only affect individuals by decreasing their sense of self-control or their subjective feelings or beliefs about support networks, but by objectively undermining the extent and effectiveness of those networks. Contexts with low overall social capital or important imbalances in the amount of bonding, bridging and linking social capital are more likely to produce, over time, widespread conditions of physical and mental mobility. Criminal networks can, through social structures, directly or indirectly affect the effectiveness of individual coping mechanisms, by modulating cognitive and emotional responses to the threat of extortion. A reasonably mentally healthy individual, for example, is unlikely to overcome extortion if governmental institutions are not trustworthy. Some contexts can also boost the effect of the threat of extortion by pairing it with additional stressors. A victim is more likely to comply with the victimiser's demand when extortion is immersed in a context of generalised violence.

3.5 Summary

The presented typology of extortion racket systems focuses on the embeddedness of extortion rackets within society and extracted the dimensions of state, criminal organisation and victims. Each major axis was divided further into two subcategories. For the state dimension these were "state responsiveness" and "dissociation for the state", for the criminal organisation "penetration of the state" and "conditions of enforcement" and for the victims they were "cognitive social capital" and "structural social capital".

The major contribution of this typology is the acknowledgement that extortion rackets cannot be understood without looking at the dimensions of the political environment they exist in (i.e. state) and the social environment (i.e. civil society, which comprises potential and actual victims). Systemic extortion is a highly complex adaptive response, associated with interaction of different actors over time. Trying to analyse ERSs devoid of the context will not lead to understanding. This holistic view on ERSs in society results in the consideration of a number of relevant aspects hitherto not collectively analysed with ERSs.

3.5.1 Socio-Spatial Factors

Although spatial aspects of extortion have been considered, they have largely been seen as territorial power dynamics where space has been conceived as static. In the analysis presented here, a critical account of how social and physical conceptions of space are partly derived from the production and reproduction of power dynamics is developed. However, control of social and physical spaces can be achieved, not by taking advantage of a power void, but by shaping the power structure in a way that favours the enforcement of extortion agreements within a single space or by the willing fragmentation of spaces.

3.5.2 Socio-Normative Factors

By contextualising the existence of extortion rackets, the role of the different actors constitutes a normative landscape with an interplay of the particular beliefs and attitudes of the participating social actors regarding deviance, crime and victimisation. These beliefs and attitudes are affected by individual mediators, such as fear of crime, and structural mediators, such as ethnicity. They are equally affected by the level of institutional penetration, directly influencing the amount and character of the resources different social actors have to invest in order to enforce or repel extortion agreements. In most contexts where systemic extortion is widespread, there are noticeable overlaps in the interests of the participating social actors. These overlapping interests reduce resource consumption, making the enforcing of the extortion relationship tenable in the long run.

3.5.3 Social-Cognitive Factors

Although there has been a focus on victims of extortion in order to measure the extent of extortion in surveys (cf. CENSIS, 2003; SOS Impresa, 2007; Euripsped, 2007), the role of the victims has been underestimated in extortion research. The typology points to mental health issues associated with extortion. Systemic extortion is likely to increase mental morbidity by producing in the victim a sense of powerlessness and hopelessness and by preventing or hindering access to social support networks. Systemic extortion also affects the dimension and effectiveness of social support networks through the systematic use of violence and by implementing social conditions that hinder solidarity and social cohesion. This aspect has an interesting reverse manifestation in the rise of civic movements against extortion rackets, such as *addiopizzo* in Italy, which seem to unite entrepreneurs in such a way that resistance is continuously strengthened.

References

Abadinsky, H. (1983). *The criminal elite*. London: Greenwood Press.
Almedom, A. (2005). Social capital and mental health: An interdisciplinary review of primary evidence. *Social Science and Medicine, 61*(5), 943–964.
Anderson, R. (1965). From mafia to Cosa Nostra. *American Journal of Sociology, 71*(3), 302–310.
Anzola, D. & Elsenbroich, C. (2015). A Typology of Extortion Racket Systems. Deliverable 1.3, Global Dynamics of Extortion Racket Systems. Retrieved from http://www.gloders.eu/images/Deliverables/GLODERS_D1-3.pdf
Briceno, F., & Bajak, F. (2015). Peru's Cocaine Backpackers Bear The Risks Of A Flourishing Trade. The World Post. Retrieved from http://www.huffingtonpost.com/2015/05/07/peru-cocaine-backpackers-photos_n_7231696.html
Bruckert, C., & Hannem, S. (2013). Rethinking the Prostitution Debates: Transcending Structural Stigma in Systemic Responses to Sex Work. Canadian Journal of Law and Society / Revue Canadienne Droit et Société, 28(1), 43–63.
Campbell, C., & Deacon, H. (2006). Unravelling the Contexts of Stigma: From Internalisation to Resistance to Change. Journal of Community & Applied Social Psychology, 16(6), 411–417.
Carr, D., & Umberson, D. (2013). The social psychology of stress, health, and coping. In J. DeLamater & A. Ward (Eds.), *Handbook of social psychology*. Berlin: Springer..
Censis-BNC Foundation. (2003). Impresa e criminalità nel Mezzogiorno. Meccanismi di distorsione del mercato. Rapporto di Ricerca. Roma.
Chang, J., Dado, D., Hawker, L., Cluss, P., Buranosky, R., Slagel, L., et al. (2010). Understanding turning points in intimate partner violence: Factors and circumstances leading women victims toward change. *Journal of Women's Health, 19*(2), 251–259.
Chin, K. (2000). Chinatown Gangs: Extortion, Enterprise, and Ethnicity. Oxford: Oxford University Press.
CNEL/ISTAT. (2014). *BES 2014*. Rome: ISTAT.
Cowen, D., & Gilbert, E. (2008). Fear and the Familial in the US War on Terror. In R. Pain & S. Smith (Eds.), Fear: Critical Geopolitics and Everyday Life. Aldershot: Ashgate.
Coleman, J. (1990). *Foundations of social theory*. Cambridge, MA: Belknap.
Decker, S., Bynum, T., & Weisel, D. (1998). A tale of two cities: Gangs as organized crime groups. *Justice Quarterly, 15*(3), 395–425.
Dickie, J. (2008). *Cosa Nostra*. London: Hodder.
EFE. (2014). Arrestan a 16 Policías Acusados de Extorsionar a Narcotraficantes en Brasil. *Página Siete*. Retrieved from http://www.paginasiete.bo/planeta/2014/10/9/arrestan-policias-acusados-extorsionar-narcotraficantes-brasil-34682.html
EFE. (2015). Expulsan a 43 Policías de Río de Janeiro Acusados de Extorsión. *Caracol Radio*. Retrieved from http://www.caracol.com.co/noticias/internacionales/expulsan-a-43-agentes-de--la-policia-de-rio-de-janeiro-acusados-de-extorsion/16173/nota/2642557.aspx
Elden, S. (2013). *The birth of territory*. Chicago: The University of Chicago Press..
Eurispes. (2007). Rapporto Italia 2007. Retrieved from http://www.eurispes.eu/content/rapporto-italia-2007.
Farley, M., & Barkan, H. (1998). Prostitution, Violence, and Posttraumatic Stress Disorder. Women & Health, 27(3), 37–49.
Fassaert, T., de Wit, M., Tuinebreijer, W., Knipscheer, J., Verhoeff, A., Beekman, A., et al. (2011). Acculturation and psychological distress among non-Western Muslim migrants—A population-based survey. *The International Journal of Social Psychiatry, 57*(2), 132–143.
Felbab-Brown, V. (2011). *Bringing the state to the slum*. BROOKINGS. Retrieved from http://www.brookings.edu/~/media/research/files/papers/2011/12/05 latin america slums felbab-brown/1205_latin_america_slums_felbabbrown.pdf
Felson, R., Messner, S., Hoskin, A., & Deane, G. (2002). Reasons for reporting and not reporting domestic violence to the police. *Criminology, 40*(3), 617–648.

Ferro, J., & Uribe, G. (2002). *El Orden de la Guerra*. Bogotá: CEJA.
Fischbach, R., & Herbert, B. (1997). Domestic violence and mental health: Correlates and conundrums within and across cultures. *Social Science and Medicine, 45*(8), 1161–1176.
Foucault, M. (1982). The subject and power. *Critical Inquiry, 8*(4), 777.
Friedrichs, D. (2010). *Trusted criminals*. London: Wadsworth.
Fugate, M., Landis, L., Riordan, K., Naureckas, S., & Engel, B. (2005). Barriers to domestic violence help seeking: Implications for intervention. *Violence Against Women, 11*(3), 290–310.
Gambetta, D. (1990). Mafia: The price of distrust. In D. Gambetta (Ed.), *Trust*. Oxford: Basil Blackwell.
Garay-Salamanca, L., & Salcedo-Albarán, E. (2012). Institutional impact of criminal networks in Colombia and Mexico. *Crime, Law and Social Change, 57*(2), 177–194.
Gieryn, T. (2000). A space for place in sociology. *Annual Review of Sociology, 26*, 463–496.
Harpham, T., Grant, E., & Rodriguez, C. (2004). Mental health and social capital in Cali, Colombia. *Social Science & Medicine, 58*(11), 2267–2277.
Harpham, T., Grant, E., & Thomas, E. (2002). Measuring social capital within health surveys: Key issues. *Health Policy and Planning, 17*(1), 106–111.
Hellman, J., Jones, G., & Kaufmann, D. (2003). Seize the state, seize the day: State capture and influence in transition economies. *Journal of Comparative Economics, 31*(4), 751–773.
Hill, P. (2006). *The Japanese mafia*. Oxford: Oxford University Press.
Hill, P. (2014). The Japanese Yakuza. In L. Paoli (Ed.), *The Oxford handbook of organized crime*. Oxford: Oxford University Press.
Ho, G. (2014). Acculturation and its Implications on Parenting for Chinese immigrants: A Systematic Review. *Journal of Transcultural Nursing, 25*(2), 145–58.
Holden, M. (2006). Exclusion, Inclusion, and Political Institutions. In R. Rhodes, S. Binder, & B. Rockman (Eds.), *Tho Oxford Handbook of Political Institutions*. Oxford: Oxford University Press.
Human Rights Watch. (2000). *Owed Justice*. New York: Human Right Watch.
Ishino, I. (1953). The Oyabun-Kobun: A Japanese ritual Kinship institution. *American Anthropologist, 55*(5), 695–707.
Jackson, J. (2008). Bridging the social and the psychological in the fear of crime. In M. Lee & S. Farrall (Eds.), *Fear of crime: Critical voices in an age of anxiety*. New York: Glass House Press.
Karstedt, S. (2014). Organizing crime: The state as agent. In L. Paoli (Ed.), *The Oxford handbook of organized crime*. Oxford: Oxford University Press.
Kawachi, I., Kennedy, B., Lochner, K., & Prothrow-Stith, D. (1997). Social capital, income inequality, and mortality. *American Journal of Public Health, 87*, 1491–1498.
Kennedy, A., Bybee, D., & Greeson, M. (2014). Examining cumulative victimization, community violence exposure, and stigma as contributors to PTSD symptoms among high-risk young women. *The American Journal of Orthopsychiatry, 84*(3), 284–294.
Killias, M. (1990). Vulnerability: Towards a better understanding of a key variable in the genesis of fear of crime. *Violence and Victims, 5*(2), 97–108.
La Spina, A. (2008). Recent anti-mafia strategies: The Italian experience. In D. Siegel & H. Nelen (Eds.), *Organized crime: Culture, markets and policies*. Berlin: Springer.
Lawler, E. (1992). Power processes in bargaining. *The Sociological Quarterly, 33*(1), 17–34.
Levada Analytical Center. (2013). *Russian public opinion 2012–2013*. Retrieved from http://www.levada.ru/sites/default/files/2012_eng.pdf
Levi, M. (2014). Money laundering. In L. Paoli (Ed.), *The Oxford handbook of organized crime*. Oxford: Oxford University Press.
Levitsky, S., & Way, L. (2010). *Competitive authoritarianism*. Cambridge: Cambridge University Press.
Loewenstein, G., Weber, E., Hsee, C., & Welch, N. (2001). Risk as feelings. *Psychological Bulletin, 127*(2), 267–286.
Lynch, J., & Kaplan, G. (1997). Understanding how inequality in the distribution of income affects health. *Journal of Health Psychology, 2*(3), 297–314.

McDermott, J. (2014). *The FARC and the Drug Trade: Siamese Twins? In SightCrime*. Retrieved from http://www.insightcrime.org/investigations/farc-and-drug-trade-siamese-twins
Malone, B. (2015). Why Al Jazeera will not say Mediterranean "migrants." *Al Jazeera*. Retrieved from http://www.aljazeera.com/blogs/editors-blog/2015/08/al-jazeera-mediterranean-migrants-150820082226309.html
Marroco, E., Fisher, J., & Kishida, I. (2008). Organised crime in the city. *Law and Financial Markets Review, 2*(4), 311–320.
Muzulu, P. (2014). Mbare Veg Market Generates US$36m. *The Standard*. Retrieved from http://www.thestandard.co.zw/2014/08/10/mbare-veg-market-generates-us36m/
Nelken, D. (2012). White-collar and corporate crime. In M. Maguire, R. Morgan, & R. Reiner (Eds.), *The Oxford handbook of criminology*. Oxford: Oxford University Press.
Pagden, A. (1990). The destruction of trust and its economic consequences in the case of eighteenth-century Naples. In D. Gambetta (Ed.), *Trust*. Oxford: Basil Blackwell.
Paoli, L. (2003). *Mafia brotherhoods*. Oxford: Oxford University Press.
Paoli, L., & Vander Berken, T. (2014). Organised crime: A contested concept. In L. Paoli (Ed.), *The Oxford handbook of organized crime*. Oxford: Oxford University Press.
Pearlin, L., Menaghan, E., Lieberman, M., & Mullan, J. (1981). The stress process. *Journal of Health and Social Behavior, 22*(4), 337–356.
Phinney, J. (1990). Ethnic Identity in Adolescents and Adults: Review of Research. *Psychological Bulletin, 108*(3), 499–514.
Putnam, R. (2000). *Bowling alone*. New York: Simon & Schuster.
Putnam, R., Leonardi, R., & Nonetti, R. (1993). *Making democracy work*. Princeton: Princeton University Press.
Ross, C., & Jang, S. (2000). Neighborhood disorder, fear, and mistrust: The buffering role of social ties with neighbors. *American Journal of Community Psychology, 28*(4), 401–420.
Ross, C., & Mirowsky, J. (2006). Social structure and psychological functioning: Distress, perceived control, and trust. In J. DeLamater (Ed.), *Handbook of social psychology*. Berlin: Springer.
Schnittker, J. (2002). Acculturation in Context: The Self-Esteem of Chinese Immigrants. *Social Psychology Quarterly, 65*(1), 56–76.
Skogan, W., & Maxfield, M. (1981). *Coping with crime*. London: Sage.
Solt, F. (2008). Economic Inequality and Democratic Political Engagement. *American Journal of Political Science, 52*(1), 48–60.
Solt, F. (2012). The social origins of authoritarianism. *Political Research Quarterly, 65*(4), 703–713.
SOS Impresa. (2007). Le mani della criminalità sulle imprese. Retrieved from http://www.sosimpresa.it/1090/le-mani-della-criminalita-sulle-imprese---xiii-rapporto-di-sos-impresa.html
Stafford, M., Chandola, T., & Marmot, M. (2007). Association between fear of crime and mental health and physical functioning. *American Journal of Public Health, 97*(11), 2076–2081.
Stenner, K. (2005). *The authoritarian dynamic*. Cambridge: Cambridge University Press.
Sung, H.-E. (2004). State failure, economic failure, and predatory organized crime: A comparative analysis. *Journal of Research in Crime and Delinquency, 41*(2), 111–129.
Szreter, S., & Woolcock, M. (2004). Health by association? Social capital, social theory, and the political economy of public health. *International Journal of Epidemiology, 33*(4), 650–667.
Taylor, B. (2011). *State building in Putin's Russia: Policing and coercion after communism*. Cambridge: Cambridge University Press.
The World Bank. (2014). *World development indicators*. Retrieved from http://data.worldbank.org/sites/default/files/wdi-2014-book.pdf
Thrift, N. (1996). *Spatial formations*. London: Sage.
Tileaga, C. (2006). Representing the "Other": A discursive analysis of prejudice and moral exclusion in talk about romanies. *Journal of Community and Applied Social Psychology, 16*(1), 19–41.
UNODC. (2010). *The globalization of crime: A transnational organized crime threat assessment*. UNODC. Retrieved from https://www.unodc.org/documents/data-and-analysis/tocta/TOCTA_Report_2010_low_res.pdf

van Dijk, T. (2012). Structures of discourse and structures of power. In J. Anderson (Ed.), *Communication Yearbook 12*. London: Routledge.
Varese, F. (2014). Protection and extortion. In L. Paoli (Ed.), *The Oxford handbook of organized crime*. Oxford: Oxford University Press.
Villarreal, A., & Silva, B. (2006). Social cohesion, criminal victimization and perceived risk of crime in Brazilian neighborhoods. *Social Forces, 84*(3), 1725–1753.
Volkov, V. (2014). The Russian mafia: Rise and extinction. In L. Paoli (Ed.), *The Oxford handbook of organized crime*. Oxford: Oxford University Press.
Wilkinson, R. (1992). Income distribution and life expectancy. *British Medical Journal, 304*(6820), 165–168.
Wright Mills, C. (2000). *The power elite*. Oxford: Oxford University Press.
Yamawaki, N., Ochoa-Shipp, M., Pulsipher, C., Harlos, A., & Swindler, S. (2012). Perceptions of domestic violence: The effects of domestic violence myths, victim's relationship with her abuser, and the decision to return to her abuser. *Journal of Interpersonal Violence, 27*(16), 3195–3212.

Part II
Society and the State

Chapter 4
Social Norms and Extortion Rackets

Áron Székely, Giulia Andrighetto, and Luis G. Nardin

4.1 Introduction

Norms, in their many forms, are all around us. The social world that we inhabit is saturated with them. As others have elegantly put it, 'from forceps to grave, human life is wrapped in a tightly woven tapestry of rules, standards, and expectations that govern every aspect of social behaviour' (Anderson & Dunning, 2014). Not only are norms widespread, but they also have powerful effects on our individual and social behaviour, often for the better but sometimes for the worse (Ellickson, 1991; Fehr & Fischbacher, 2004; Henrich et al., 2001).

Actions taken within the context of protection rackets are arguably not exempt. Domain-specific norms, regarding paying or supporting behaviour, and domain-general

Á. Székely
Institute of Cognitive Sciences and Technologies, Italian National Research Council (CNR), Via San Martino della Battaglia 44, Rome 00185, Italy
e-mail: aron.szekely@istc.cnr.it

G. Andrighetto (✉)
Institute of Cognitive Sciences and Technologies, Italian National Research Council (CNR), Rome, Italy

Schuman Centre for Advanced Studies, European University Institute, Fiesole, Italy
e-mail: giulia.andrighetto@istc.cnr.it

L.G. Nardin
Institute of Cognitive Sciences and Technologies (ISTC), Italian National Research Council (CNR), Via Palestro, 32, Rome 00185, Italy

Schuman Centre for Advanced Studies, European University Institute, Fiesole, Italy
e-mail: gnardin@gmail.com

C. Elsenbroich et al. (eds.), *Social Dimensions of Organised Crime*, Computational Social Sciences, DOI 10.1007/978-3-319-45169-5_4

norms such as reciprocity and fairness may shape behaviour there too. Moving from the general commitment to including norms in a model of protection rackets to their implementation is complex and multilayered; this chapter helps lay the foundations. Specifically we address:

- Basic definitions of norms
- Important empirical features of norms
- Key norms in protection rackets
- Interactions between protection racket-related norms

In the past, the concept of norms was rightly criticised for being vague and operating behind the back of their subjects ignoring people's preferences and decisions (e.g. Gambetta, 1993, pp. 10–11). However, more recent theory specifies them precisely in ways that are compatible with a choice-based perspective of human behaviour (e.g. Bicchieri, 2006; Crawford & Ostrom, 1995; Elster, 2007). A key point of such theories is that norm compliance is *conditional*, rather than an unconditional commitment to follow a rule; how people behave is affected by the beliefs that they have about the expectations of others and the beliefs they have about their potential to be sanctioned for deviating. Experiments, for instance, show that changing these expectations affects behaviour (e.g. Bicchieri & Chavez, 2010; Bicchieri & Xiao, 2009). Neither do these theories imply that social norm compliance is irrational: given certain preferences and expectations it is rational to comply.

4.2 Norm Fundamentals

There is consensus among social scientists that norms, in one form or another, are important factors affecting human behaviour (Opp & Hechter, 2005; Xenitidou & Edmonds, 2014). While there is widespread agreement about the important role that norms play in social life, there is much disagreement about their *specifics* regarding conceptualisation and definition (e.g. Bicchieri, 2006; Cialdini, Reno, & Kallgren, 1990; Coleman, 1990; Conte, Andrighetto, & Campennì, 2013; Crawford & Ostrom, 1995; Elster, 2009; Finnemore & Sikkink, 1998).

In everyday terminology, a 'norm' refers to something that is usual, typical or standard:

> 'a standard or pattern, especially of social behaviour, that is typical or expected'; or
>
> 'a required standard; a level to be complied with or reached' (*Oxford Dictionary of English*, 2012).

According to this definition, a 'norm' can refer to a standard pattern or a required standard. The issue with this basic definition is that it confuses 'both behaviour that *is* normal, and behaviour that people *should* mimic to avoid being punished' (Ellickson, 1991, p. 126). It muddles the important descriptive versus prescriptive distinction. Clearly, the everyday meaning of norms refers to an umbrella of social phenomena that we must carefully define and separate.

Out of these phenomena, we can distinguish and define two important kinds of norms: *social norms* and *legal norms*. These two concepts have broadly agreed-upon definitions in the literature, albeit some people use different terms for naming them, and they are the types of norms that are explicitly, or implicitly, included in virtually every influential theory of norms.

Social norms can be defined as shared behavioural rules that proscribe or prescribe ways of acting that are followed because of reciprocal expectations and in some cases social punishment (Bicchieri, 2006; Cialdini et al., 1990; Elster, 1989).

From the decades of observational studies, we know that social norms affect how people behave in innumerable contexts. Ellickson (1991), for instance, in his famous book, *Order without Law*, describes the social norms regulating cattle trespassing, which apply more broadly to ownership and liability, in Shasta county. He compares how conflicts are resolved and cooperation achieved in two different types of ranges, 'open' and 'closed', in which cattle owners' liabilities, according to legislation, are completely different. One of his key findings is that in many situations, residents avoid the law and do not respond to the legal distinctions, and instead 'apply informal norms, rather than formal legal rules, to resolve most of the issues that arise among them' (p. 1). By relying on social norms to resolve conflict, costly legal wrangles are avoided.

Experiments have allowed researchers to study the causal effect of social norms on behaviour. In a classic study from social psychology, Cialdini and colleagues carried out five experiments on norms. In one, in which they specifically test social norms (which they refer to as 'injunctive norms'),[1] they manipulated (1) the perceived appropriateness of the social norm of littering by sweeping trash into a pile or leaving it strewn about and (2) subjects' attention to this norm cue by having a confederate litter, or not, in front of them (1990, pp. 1020–1022). They found that when subjects had their attention drawn to the norm cue, and the cue indicated that it was inappropriate to litter, substantially more subjects avoided littering—conforming to the norm—than when the cue indicated that it was not problematic to litter. In contrast, when subjects did not have their attention drawn to the norm, their behaviour did not differ according to the cue. Thus, being made aware of a norm led to changes in behaviour while without awareness no difference was observed.

Multiple other experiments have now also demonstrated social norms' effects on behaviour in laboratory settings (e.g. Aarts & Dijksterhuis, 2003; Bicchieri & Chavez, 2010; Krupka & Weber, 2009) and in the field (Allcott, 2011; Hallsworth et al., 2016; UK Behavioural Insights Team, 2013). Other experiments have begun to identify the neurological underpinnings of social norm compliance (Knoch, Pascual-Leone, Meyer, Treyer, & Fehr, 2006; Ruff, Ugazio, & Fehr, 2013), and at least one field experiment demonstrates that social norms can cross-contaminate such that compliance and violation with one social norm promotes or inhibits compliance with another, related, social norm (Keizer, Lindenberg, & Steg, 2008).

[1] Cialdini et al. (1990) refer to 'social norms' as 'injunctive norms'; the two are essentially identical. Researchers in social psychology, drawing on the work of Cialdini and colleagues, typically use the term injunctive norms (e.g. Anderson & Dunning, 2014).

Legal norms can be considered as shared behavioural rules that are formalised as laws and enforced by specialised actors (Bicchieri, 2006; Elster, 2009).

Laws make the shared behaviour rule component of a legal norm explicit and legally enforceable. Often enforcement is carried out by punitive means: some sort of cost, a fine or imprisonment, for instance, is imposed on the lawbreaker.

To highlight, the analytic distinction between social and legal norms is that the former are only ever socially enforced, if at all, while the latter are enforced by governmental law enforcement actors, the police, for instance, who are responsible for carrying out this process. (See Chap. 5 for a detailed analysis of legal responses to the extortion phenomenon in Italy.)

Both social and legal norms can exist without being actively followed. As long as enough people in the population know about the behavioural rule and a large enough proportion of them would follow the rule if others were to also do so, or if the sanction were present, then it exists in a latent form. When people start conforming to the shared behavioural rule, it becomes active (Bicchieri, 2006, p. 11). Thus, the observation that people are not taking an action is not evidence that there is no social or legal norm regulating that behaviour.

Two further relevant points should be addressed before proceeding. First, as Elster writes, while the two concepts are analytically separable, in practice the same behaviour may be associated with multiple norms (2009, p. 198). Behaviour *X* may be both a social norm and a legal norm; smoking in enclosed places is illegal in Britain since 2006 and 2007 but is now also a social norm—a few months ago one of the authors witnessed a traveller waiting for a train chastising another for smoking at the station platform. Second, and related to the prior point, it is difficult to tell exactly what is driving a behaviour. Avoiding the behaviour *Y* may be motivated by a social norm, a legal norm, personal beliefs, emotions or incentives, among others. Similarly, some behaviours are somewhere in between social and legal norms: formally enforced rules in a company for instance. Recent work has started to develop toolkits that allow us to measure and distinguish whether norms, and which ones, are motives for a behaviour (Bicchieri, Lindemans, & Jiang, 2014). Without experimental causal testing that goes into determining whether a norm affects or not a behaviour (e.g. Horne, Dodoo, & Dodoo, 2013) we cannot know for sure whether a norm drives a behaviour; detailed empirical, both experimental and non-experimental, tests of norms related to protection rackets should be a priority for future work.

4.3 Norms in Protection Rackets

Norms are often specific to the protection racket and the context within which they operate. Our focus is on the Sicilian mafia and the norms that we discuss concerning protection rackets are drawn from this case. Some may also apply to other protection rackets while others may not. We identified the specific social and legal norms relevant to the Sicilian protection racket based on discussions with the GLODERS project members; discussions with the project stakeholders; work compiled by Militello, La Spina, Frazzica, Punzo, and Scaglione (2014), reviewing the primary and secondary literature; and discussions with an Addiopizzo committee member.

4.3.1 Legal and Social Norms in Protection Rackets

Italian legal norms that specifically apply to protection rackets can be broadly split into those that combat mafia and allow the confiscation of their resources and those that help affected shopkeepers and business people.

Considering briefly some of the important legislation, the *Rognoni-La Torre Law n. 646 of 13/9/1982* introduced into the Italian criminal code the crime of mafia-style criminal organisation (art. 416 bis) and the possibility of confiscating mafia properties with their consequent social reuse. In addition, *Law n. 8 of 15/01/1991* and *Law n. 82 of 15/03/1991* aim at providing reporting incentives and protecting victims who report extortion activities. Finally, *Law n. 44 of 23/02/1999* and *Law n. 512 of 22/12/1999*, respectively, introduced economic support to victims of extortions and the solidarity fund for victims of mafioso-style crimes and intimidation. For further details on the regulatory response to the mafia see Chap. 5.

Turning to norms that are not backed by legislation, i.e. social norms, a number may be at play. Reciprocity norms between shopkeepers who pay pizzo and mafiosi, who are then expected to give something in return, may play a part, and 'fairness' too, similarly between mafiosi and shopkeepers; in a recent laboratory experiment Andrighetto and co-authors found that subjects participating as 'mafiosi' request a fair amount of pizzo—protection money paid to mafia—from 'shopkeepers', with inflicted punishment being rare and the sums requested being small (2015). Paying pizzo becomes normalised among business people as it is turned into an ordinary component of the market (Asmundo & Lisciandra, 2008). Social punishment for reporting: in one case, citizens boycotted a shopkeeper because he reported pizzo requests to the police (Diliberto, 2013). Codes concerning the divulgence of information, *omertà*, to authorities are another; secrecy clearly can help the mafia. As Vaccaro also writes 'Sicilians struggle to understand that (a) colluding with the mafia is not ethically proper' (p. 29). This is partly what Addiopizzo, an anti-mafia non-governmental organisation (NGO), targets: they use a variety of strategies (more on this later) to try and make it socially and morally unacceptable to pay. In-group favouritism that is socially triggered, when observable and punishable by others, could play a role. One study that looks at the behaviour of school children in an experimental setting who come from high- and low-mafia-activity neighbourhoods finds evidence for this (Meier, Pierce, & Vaccaro, 2013). They find in-group favouritism among school children from a high-mafia-presence area but not among those from a low-mafia-presence area and only when a third party observed, and could punish, their behaviour.

Regarding social norms that apply to consumers' decisions, there do not seem to be any that currently broadly apply. A small proportion of Sicilians may be affected by an 'emerging' social norm to avoid shops that pay pizzo—a social norm that NGOs are specifically targeting. Traditionally, as well as currently, this has not been a prevailing social norm among people in Sicily. Certainly, if all shopkeepers pay pizzo, then consumers have no choice in this regard, and so, it is difficult to comply and establish this norm. Recent efforts by non-governmental organisations, in particular Addiopizzo, have focused on encouraging this social norm. One way in which they do so is with a 'sticker strategy'. Entrepreneurs can apply to Addiopizzo

Table 4.1 A summary of the implemented norms

Legal norms	Social norms
Criminalises mafia-style organisations (*Law n. 646 of 13/9/1982).*	*Reciprocity and fairness* between pizzo payer and racketeer
Reporting incentives and protection (*Law n. 8 of 15/01/1991* and *Law n. 82 of 15/03/1991*).	*Pay pizzo* among shopkeepers
Economic support to extortion victims (*Law n. 44 of 23/02/1999* and *Law n. 512 of 22/12/1999*)	*Not pay pizzo* among shopkeepers
	Not report paying to the authorities
	Report paying to the authorities

to receive a certification of non-payment that they can display to potential customers in the form of a highly visible sticker attachable to the shop front window. Addiopizzo thoroughly vets businesspeople who apply to ensure that only genuine non-payers receive the displayable signal. This helps to solve the issue of choice. Another strategy is more direct: Addiopizzo collects signatures from consumers who declare that they will not buy products from shops that pay pizzo. The hope is that this will serve as a commitment device and foster 'critical consumers' who follow this norm.

The state, mafia, and non-governmental organisations also have social norms within them that affect how people behave and the functioning of the institutions. Codes of conduct are codified within the Sicilian mafia (Gambetta, 1993)—Pizzini-Gambetta (1999), for instance, highlights and discusses some of the gender norms in the mafia and their strategic exploitation also. In contrast to the approach taken in the 'Palermo Scenario', a specific implementation of a protection racket in Palermo (see Chap. 7 for details), including such intra-institutional norm dynamics would necessitate a greater focus on each of these actors, as opposed to the inter-actor dynamic, that the Palermo Scenario undertakes. See Table 4.1 for a summary of the legal and social norms that we identified.

4.3.2 *Other Norms in Protection Rackets*

In addition to social and legal norms scholars distinguish between a variety of other concepts, including descriptive norms, moral norms, quasi-moral norms, and conventions that overlap in various degrees and in various ways with certain definitions of social and legal norms.

Our focus is on legal and social norms for a number of reasons. First, they seem to play the most important roles in determining the success and failure of protection rackets. Legislation is the prime tool employed by governments to destabilise protection rackets, physically constraining the behaviour of people and changing the anticipated costs to taking an action. Social norms were identified, in part, as important based on discussions with the project stakeholders, an Addiopizzo board member, and discussions with University of Palermo unit in the GLODERS project. Second, both legal and social norms, to a degree, are changeable by people who come from government departments, the judiciary, and NGOs. Consequently, focusing on legislation and social norms enables us to understand their consequences and present the result of tools to a broad range of potential beneficiaries who might then use them.

Moral norms are the other type of norms that would, to a degree, also satisfy the conditions set out above. However, in some definitions of moral norms, they are unconditional, and very difficult to change; people who possess those moral norms will follow them, more or less irrespective of what others do and expect (e.g. Bicchieri, 2006; Elster, 2009).

4.4 Interdependence of Legal and Social Norms

So far, legal and social norms have been discussed in isolation. Yet, since legal norms are 'always imposed against a background stream of nonlegal regulation' (Posner, 2002, p. 4), social norms emerge against a backdrop of legal norms, and they affect one another, it is important to consider some ways in which they interact. We turn to this topic now.

Social norms are often implicitly or explicitly considered to be socially and individually beneficial (e.g. Ellickson, 1991).[2] In many cases this is a correct assumption; social norms can increase cooperation, facilitate economic exchange, and allow disputes to be settled without a reliance on costly legal intervention (Ellickson, 1991; Fehr & Fischbacher, 2004; Ostrom, 2000). Yet, other times, social norms appear to be harmful.[3] Norms of honour encourage 'honour' killings (e.g. Mojab & Hassanpour, 2003) and contribute to perpetuating cycles of revenge. Norms concerning marriage, in some contexts, have led to foot binding and infibulation (Mackie, 1996). In our context, social norms for supporting, or at least not hindering, mafiosi, and omertà, may be important for the success of protection rackets.

Bottom-up actions, carried out by individuals who have no legal enforcement capabilities, can contribute to the change and collapse of harmful actions that are supported by social norms. As argued forcefully by Gerry Mackie, peer effects are the key to reducing female genital cutting (Mackie, 1996, 2000). There is some evidence that women in Senegal who were supported by the NGO Tostan have been able to substan-

[2] Although Ellickson defines social norms as 'welfare maximising' (1991, p. 167), and thus fits this consideration, his definition is far from a naïve functionalist one. His definition is, for instance, compatible with social norms that at first glance appear to contradict it (see Chap. 15).

[3] Clearly, it is not straightforward to define what is a harmful social norm and what is not. This could depend on whether one subscribes to, for instance, a utilitarian view of ethics. Nevertheless, the examples presented have a strong element of harm to them.

tially reduce female genital mutilation and cutting in their villages (UNICEF, 2008). Education programmes combined with public declarations have led to the wholesale abandonment of the practice in some communities. One village chief describes how 'a broad discussion took place, largely conducted by the women themselves. All of us accepted to abandon this custom, which has nothing Islamic about it. This commitment we made in front of the whole world' (UNICEF, 2008, p. 35). The commitment broke the social norm so that deviating from the practice was no longer associated with negative individual repercussions, and consequently, people stopped the practice of female circumcision.[4] Among the Acholi, in Uganda, some individuals contest existing social norms that legitimise domestic violence by referring to the 'Acholi cultural traditions in order to build more peaceful communities' (Lundgren & Adams, 2014, p. 53). They try to break down the social norm of 'manliness' that leads to domestic violence by reframing what manliness should consist of.

Yet, bottom-up actions may not be enough to displace a social norm. Being a first-mover can have substantial costs associated with it since it leaves the initiating individual vulnerable to punishments from others. In addition, endogenously emerging social norms may end up supporting new actions that are harmful to a society; unguided, they can lead to damage.

Legal norms can be used as a top-down method to influence behaviour and social norms to push them out of equilibrium and guide them in specific directions. Put broadly, legal norms can affect social norms in two ways: (1) legal norms can contribute to the *breakdown* of social norms, and (2) legal norms can *build up*, or guide, social norms. The landmark case of Brown vs. Board of Education (1954) may be an example of a legal decision that led directly to the desegregation of schools and arguably indirectly to the breakdown, or reduction, in discriminatory social norms. An example of the latter—legal norms building up social norms—can be found in the internal policies of North Korea. The regime's policy of monitoring the population and severely punishing offenders, sending minor rule-breakers for 're-education' and more serious offenders to internment in political prison camps, is a powerful tool that directly and indirectly enables the state to build leader worship and a social norm of state support (Byman & Lind, 2010). Without the powerful coercive measures, propaganda and the co-option of 'troublesome' classes may not be enough.

Legislation alone is a powerful tool that prevents many poor outcomes from occurring, among them anti-discrimination laws, property rights, and traffic laws. However, by themselves, laws are not always sufficient to produce a desired behaviour; a brief glance at the state of parking on many streets in Rome provides ample evidence of the former: while regulation prohibits parking in certain places it is clear that some people do not conform to it. Other actions lie outside its influence because of the difficulties in their monitoring and enforcement. Even when technically possible, relying on enforcement is costly and inefficient and behaviour that

[4] Recent work however argues for a primarily peer convention account, as opposed to marriage market convention account, of female genital cutting (Shell-Duncan, Wander, Hernlund, & Moreau, 2011) and another questions the entire convention aspect involved (Efferson, Vogt, Elhadi, Ahmed, & Fehr, 2015).

rests solely on legal power is brittle and can collapse when enforcement subsides (Tyler, 2004, 2006).

Social norms can affect legal norms in two main ways. (1) Social norms can *undermine* legal norms when the two conflict. Omertà and support for the mafia prevent pizzo requests and punishments from being reported, undermining anti-mafia efforts. (2) Social norms can *support* legal norms when they are in harmony. Since social norms can support and supplement legal norms, understanding how they interact is crucial.

When social norms and legal norms are congruent, decentralised social enforcement becomes a factor that supports compliance with the law. Moreover, lawbreaking, in this case, would, at least some of the time, be accompanied by reporting to the state. Both factors make law-abiding behaviour likelier since the probability of enforcement increases. Compliance with the law, based on a preference to comply with the norm—even in the absence of punishments, would support legal norm compliance too. Finally, social norm support should imbue legislation with greater resilience to a lack of formal enforcement; even when the state cannot punish lawbreaking, if that law is supported by a social norm, social punishment may be enough to obtain compliance.

4.5 Norm Change and Protection Rackets

4.5.1 The Sicilian Mafia

The birth of the Sicilian mafia can be traced back to the mid-nineteenth century, in the decades following the unification of Italy (Gambetta, 1993; Pezzino, 1990). From the late nineteenth century onwards, the term mafia 'began to be restricted to those groups and even single individuals active in Sicily that systematically resorted to violence and the threat of violence, in order to control the political and economic life of their towns and villages, and whose power was usually accepted by the local population' (Paoli, 2014).

Despite its long history, the recognition of the mafia as a criminal group by the state was only recently consolidated through the introduction of the Rognoni-La Torre Law (*law n. 646 of 13/9/1982*) of art. 416 bis into the Italian Penal Code. This law regulates mafia-type criminal association (La Spina, 2014; Paoli, 2014). Since its introduction, and the evidence emerging from the Maxi Trial, the existence of the mafia as a criminal group has no longer been questioned (Catanzaro, 1988; La Spina, 2005; Paoli, 2000; Pezzino, 1990). They are now frequently defined as criminal groups that are in the business of producing, promoting, and selling protection (Gambetta, 1993), supporting the view of mafias as 'professional organisations' (Gambetta, 1993; La Spina, 2005).

The mafia acts within a vast and branched relational context and sets up a system of violence and lawlessness for the acquisition of capital and the management of power positions (Santino, 1995). However, its typical, and characteristic, criminal activity is the protection racket expressed through the mechanism of *protection-extortion* (Gambetta, 1993; La Spina, 2005): the production and sometimes forced supply of private protection in exchange for money (pizzo) or other resources, under the threat of punishment (Gambetta, 1993; La Spina, 2005; Scaglione, 2008; Schelling, 1967, 1971). Mafias provide protection services to businesses replacing the state in the resolution of many different conflicts (Gambetta, 1993; Varese, 2011, 2014). In return, they increase their authority and reputation within the territory in which they operate, achieving consensus on part of the society that can be used to mobilise citizens against the law enforcement (e.g. non-cooperation with police investigations). Other times the only protection provided is from the extorters themselves (Varese, 2014).

Through the *protection-extortion* mechanism, the mafia pursues two key objectives: firstly economic by the constant and regular considerable profits, and secondly criminal consisting of a systematic control of the territory in which they operate allowing them to perform other sorts of criminal activities.

This protection-extortion has recently been studied as a dynamic process (Punzo, 2016; Scaglione, 2008; Sciarrone, 2009). This perspective on extortion allows us to assess the different paths arising from the interactions among the actors involved in the extortion process and to understand the dynamics of its emergence and spreading within a territory (Sciarrone, 2009). Additionally, it has helped us to identify two potential approaches for countering these criminal groups: a *legal* approach that is already much used and a *social* norm approach that has only recently begun to emerge. The former is used to coercively reduce the power of the mafia, while the latter promotes the emergence and spread of social norms supporting a 'culture of legality'.

A number of non-governmental organisations have emerged in Sicily, among them Addiopizzo, Fondazione Rocco Chinnici, Libera, and Professionisti Liberi, that try to combat the mafia through norm and attitude change. We focus on one in particular, Addiopizzo, and examine the main social norm anti-mafia strategies that it uses.[5]

Addiopizzo, which was started in 2004 by a group of friends following a discussion about how they would react to having to pay pizzo, employs five separate and important approaches (Superti, 2009; Vaccaro, 2012, p. 28). It:

- Certifies shops as pizzo-free and provides them with a visible indicator of their certification. Shopkeepers can then place this manifestation of their non-payment in their shop front windows to serve as a reliable signal to potential shoppers and to other shopkeepers. Addiopizzo uses a rigorous vetting process to ensure that shopkeepers really do not pay, allowing the signal to maintain reliability.
- Condemns mafia activity in the media.
- Educates schoolchildren with various campaigns.
- Collects signatures from consumers in which they declare that they will avoid pizzo-paying shops, potentially functioning as a commitment device.
- Employs a policy of transparency regarding their own actions and demonstrating their trustworthiness in an attempt to gaining legitimacy among the population.

[5] A part of this information comes from the two interviews conducted with an Addiopizzo committee member.

One of their aims is to create a set of 'critical consumers' who avoid pizzo-paying shops for normative reasons, thereby putting pressure on shops to join the organisation, attain pizzo-free status, and actually stop paying pizzo. Crucial to this endeavour is the reliability of the vetting process that allows shopkeepers to attain the pizzo-free status. If too many mimics—shops paying pizzo but attaining the Addiopizzo sticker—join, the reliability of sticker-signal is undermined and potential consumers will stop paying attention to it. Such strategies by the relevant NGOs may help to build a community of citizens who are ideologically opposed to mafia and are willing to incur costly punishment to signal their commitment to the social norm of 'do not pay pizzo' to others in society.

The two-pronged approach to targeting the mafia, with both legal and cultural sides, is an approach that is advocated for among scholars by Roy Godson and co-authors (Godson, 2000; Godson, Kenney, Litvin, & Tevzadze, 2004) and by Leoluca Orlando, the current, and previous, mayor of Palermo (Orlando, 2001). They argue that the development of a 'culture of lawfulness' is a crucial factor in fighting organised crime. 'Bolstered by a sympathetic culture—culture of lawfulness—law enforcement and regulatory systems function more effectively in myriad ways. Those who transgress the rules find themselves targeted not only by law enforcement but also by many sectors of society. Community support and involvement can also focus on preventing and on rooting out criminal and corrupt practices without the need for expenditures for a massive law enforcement and punitive establishment. This involvement also reduces the risk and expense of intrusive government surveillance and regulatory practices harmful to individual liberties and creative economic, social, and political initiatives' (Godson, 2000, p. 92).

By its nature, it is difficult to know exactly how much of an effect the combined approach has had in Sicily. We have run some simulations on this that help us to disentangle their effects and combined set-ups seem to produce stronger anti-mafia outcomes. One finding that emerges is that the combination of legal and social norm approaches leads to the reduction in the strength of the mafia coupled with changes in the behaviour of shopkeepers (see Nardin et al., 2016; Troitzsch, 2015). Integrating real data with these simulations can help validate their predictions (see Chap. 12).

Since the 2000s, the Italian state has, in addition to a legislative approach, started to take a cultural or social norm approach towards reducing the influence of the mafia. It supports legality initiatives (e.g. *Festival della Legalità*) and education campaigns, and spreads information about successful anti-mafia operations through the media. We turn now to a country in which this combined approach, arguably, has had a profound effect in decreasing their own mafia.

4.5.2 *The Georgian Mafia*

Georgia is a small post-Soviet country with around five million inhabitants that gained its independence following the dissolution of the USSR in 1991. Its experience provides a useful contrasting case study to Sicily with which to explore a beneficial interaction between legal and social norms and the effects that they have on mafias in a country. Following independence, the thieves-in-law—a kind of mafiosi—and criminality were widespread; however today, both have been dramatically reduced. Georgia 'has moved from being a "failed state" with high corruption rates, a strong presence of organized crime, and with uncontrolled areas … to the state that effectively curbed the influence of criminal leaders, eliminated corruption in lower and medium ranks of bureaucracy and reclaimed the monopoly over the means of violence' (Kupatadze, 2012, p. 18).

Having lain dormant, or hidden, under pressure from Soviet policies, the Georgian thieves-in-law emerged strongly in the power vacuum that was left behind the USSR (Godson et al., 2004; Kukhianidze, 2009; Slade, 2012a, 2012b). By the end of the 1990s, a third of all thieves-in-law believed to be active in the post-Soviet countries were of Georgian origin (reported in Slade, 2012a, p. 38). Highly influential, they were connected to the government and law enforcement agencies (Godson et al., 2004; Kakachia & O'Shea, 2012).

Simultaneously, a kind of anti-legal culture could be identified among Georgians. As in Sicily, in part this culture can be considered as social norms that support or allow mafias to be left alone: among them the social monitoring and enforcement of behaviour supporting and stifling criticism of the thieves-in-law and not cooperating with authorities. Godson and co-authors argue that in 'Georgian culture, criminal figures have often been characterized as servants of the public good, embodying popular notions of honor, justice, or even democracy', generally role models to be imitated (2004, p. 11). 'The great Soviet heroes are no longer the role models for Russian youth. Instead, the stock broker and the *mafiya* hitman have replaced these icons' (DiPaola, 1996, p. 156), citing (Handelman, 1997, p. 345). In addition, Slade reports that people had 'positive normative orientations' towards the thieves-in-law (Slade, 2012b, pp. 629–630). 'Various forms of corruption, exploiting the state for individual/familial gain, were not often considered crimes but, on the contrary, legitimate and even proper, to the extent that, in 1993, 25 per cent of schoolchildren interviewed in Georgia said they wanted to be a thief-in-law when they grew up' (Kakachia & O'Shea, 2012, p. 11). Furthermore, Godson and co-authors argue that an irrelevant education system is also an issue: 'from kindergarten through university, students are socialized to see few alternatives to corruption and little incentive to obey the law' (pp. 12, 13), as well as the Orthodox Church which has tolerated 'criminal and illegal activities' (p. 14).

Writing in 1996, DiPaola took a foreboding tone about the chances of success in the recently formed post-Soviet states, including Georgia (DiPaola, 1996). He based this pessimism on the strength and professionalisation of the mafia, the lack of protection afforded by the state to its citizens, and importantly too the absence of a legal culture.

Yet by 2004 the situation had dramatically changed. Support for the thieves-in-law was low: only 12 % of respondents in a national survey viewed them favourably—the lowest of any institution that data was collected for (International Republican Institute, 2004, p. 52), and by 2007 only 7 % of respondents thought the same (International Republican Institute, 2007, p. 67). Moreover, in a 2010 survey, 86 % of respondents answered that the authority of the thieves-in-law had decreased (10 % responded decreased somewhat, 70 % responded decreased significantly, and

6 % eliminated) (reported in Slade, 2012b, p. 627). And in 2006 'the general prosecutor reported that there was not one thief-in-law left in Georgia outside prison' (quoted in Slade, 2012b, p. 627).

How did this change occur? One answer in the literature is that the Georgian Government implemented a two-pronged anti-crime and -corruption strategy after around 2003 consisting of (a) a legal aspect and (b) a cultural or social aspect (Godson et al., 2004; Slade, 2012a, 2012b; Tsitsishvili, 2010). This strategy was modelled on the Italian approach and was partly inspired by Leoluca Orlando (Slade, 2012a, p. 43). As part of the former, legislation was introduced that made being a thief-in-law illegal and 'enabled the state to confiscate property, and offered extensive protection to witnesses' (Kakachia & O'Shea, 2012). In addition, large sections of the police were fired, and rehired if suitable, their salaries were substantially increased, and relevant institutions were restructured (Kakachia & O'Shea, 2012; Kukhianidze, 2009; Slade, 2012a). Regarding the latter, a 'Culture of Lawfulness' curriculum was introduced into secondary schools (Godson et al., 2004, pp. 19–21)—in 2004 it was being taught in nearly half of the schools in Tbilisi (p. 22)—and 2008 onwards an NGO, Project Harmony, ran a civic education course (Slade, 2012a, p. 47). Such an approach may change norms as well as reduce the potential pool of new recruits among Georgian youth.

An important question remains. Why has this combined approach not worked in Sicily to the same extent as it did in Georgia? Clearly there are myriad potential answers, and the aim here is not to isolate nor argue for an explanation. Yet, one fascinating possibility, which was proposed by Slade (2012b) as a key factor in the Georgian mafia's downfall, is that by the time the policies were implemented, the thieves-in-law were already vulnerable. Increased criminal competition reduced the quality of recruits, increased opportunities to generate wealth reduced mutual monitoring, distrust in the commitment to shared norms increased, and discount rates on the value of possessing the thief-in-law name therefore increased (pp. 632–633). This then led to a reduction of support from the population and subsequently 'support for the thieves-in-law appeared to evaporate overnight with the anti-mafia reform' (p. 644). According to this line of argument, internal factors destabilised the mafia, changing attitudes and norms in the population before the top-down policies were implemented. This is in contrast to the Sicilian case in which top-down policies have been implemented without social norm change.

4.6 Summary

In this chapter, we have, drawing on the current literature, defined and discussed social and legal norms, and explored some of their differences and similarities. We then turned to an exploration of some of the legal and social norms arguably relevant in protection rackets. We also discussed some of the ways in which norms, relevant to protection rackets, interact, and draw on the Sicilian mafia and the Georgian mafia as two cases. The next chapter will focus in more detail on legal norms and their historic development in the case of the Italian mafia.

References

Aarts, H., & Dijksterhuis, A. (2003). The silence of the library: Environment, situational norm, and social behavior. *Journal of Personality and Social Psychology, 84*(1), 18–28.
Allcott, H. (2011). Social norms and energy conservation. *Journal of Public Economics, 95*(9–10), 1082–1095. http://doi.org/10.1016/j.jpubeco.2011.03.003
Anderson, J. E., & Dunning, D. (2014). Behavioral norms: Variants and their identification. *Social and Personality Psychology Compass, 8*(12), 721–738. http://doi.org/10.1111/spc3.12146
Andrighetto, G., Grieco, D., & Conte, R. (2015). *Fairness and compliance in the extortion game*. EUI Working Paper, European University Institute, Florence.
Asmundo, A., & Lisciandra, M. (2008). The cost of protection racket in Sicily. *Global Crime, 9*(3), 221–240. http://doi.org/10.1080/17440570802254338
Bicchieri, C. (2006). *The grammar of society: The nature and dynamics of social norms*. New York: Cambridge University Press.
Bicchieri, C., & Chavez, A. (2010). Behaving as expected: Public information and fairness norms. *Journal of Behavioral Decision Making, 23*(2), 161–178.
Bicchieri, C., Lindemans, J. W., & Jiang, T. (2014). A structured approach to a diagnostic of collective practices. *Cultural Psychology, 5*, 1418. http://doi.org/10.3389/fpsyg.2014.01418
Bicchieri, C., & Xiao, E. (2009). Do the right thing: But only if others do so. *Journal of Behavioral Decision Making, 22*(2), 191–208. http://doi.org/10.1002/bdm.621
Byman, D., & Lind, J. (2010). Pyongyang's survival strategy: Tools of authoritarian control in North Korea. *International Security, 35*(1), 44–74. http://doi.org/10.1162/ISEC_a_00002
Catanzaro, R. (1988). *Il delitto come impresa: Storia sociale della mafia*. Padova: Liviana.
Cialdini, R., Reno, R. R., & Kallgren, C. A. (1990). A focus theory of normative conduct: Recycling the concept of norms to reduce littering in public places. *Journal of Personality and Social Psychology, 58*(6), 1015–1026. http://doi.org/10.1037/0022-3514.58.6.1015
Coleman, J. S. (1990). *Foundations of social theory*. Cambridge, MA: Harvard University Press.
Conte, R., Andrighetto, G., & Campennì, M. (Eds.). (2013). *Minding norms: Mechanisms and dynamics of social order in agent societies*. New York: Oxford University Press.
Crawford, S. E. S., & Ostrom, E. (1995). A grammar of institutions. *American Political Science Review, 89*(3), 582–600.
Diliberto, P. (2013, April 1). Addiopizzo 2.0. MTV.
DiPaola, P. D. (1996). Criminal time bomb: An examination of the effect of the Russian Mafiya on the newly independent states of the former Soviet Union. *Indiana Journal of Global Legal Studies, 4*, 145.
Efferson, C., Vogt, S., Elhadi, A., Ahmed, H. E. F., & Fehr, E. (2015). Female genital cutting is not a social coordination norm. *Science, 349*(6255), 1446–1447. http://doi.org/10.1126/science.aaa7978
Ellickson, R. C. (1991). *Order without law: How neighbors settle disputes*. Cambridge, MA: Harvard University Press.
Elster, J. (1989). Social norms and economic theory. *The Journal of Economic Perspectives, 3*(4), 99–117.
Elster, J. (2007). *Explaining social behavior: More nuts and bolts for the social sciences* (2 Revth ed.). Cambridge: Cambridge University Press.
Elster, J. (2009). Social norms. In P. Hedström & P. Bearman (Eds.), *The Oxford handbook of analytical sociology*. Oxford: Oxford University Press.
Fehr, E., & Fischbacher, U. (2004). Social norms and human cooperation. *Trends in Cognitive Sciences, 8*(4), 185–190.
Finnemore, M., & Sikkink, K. (1998). International norm dynamics and political change. *International Organization, 52*(4), 887–917. http://doi.org/10.1162/002081898550789
Gambetta, D. (1993). *The Sicilian mafia: The business of private protection*. Cambridge, MA: Harvard University Press.

Godson, R. (2000). Guide to developing a culture of lawfulness. *Trends in Organized Crime, 5*(3), 91–102. http://doi.org/10.1007/s12117-000-1038-3

Godson, R., Kenney, D. J., Litvin, M., & Tevzadze, G. (2004). Building societal support for the rule of law in Georgia. *Trends in Organized Crime, 8*(2), 5–27. http://doi.org/10.1007/s12117-004-1007-3

Hallsworth, M., Chadborn, T., Sallis, A., Sanders, M., Berry, D., Greaves, F., et al. (2016). Provision of social norm feedback to high prescribers of antibiotics in general practice: A pragmatic national randomised controlled trial. *The Lancet*. http://dx.doi.org/10.1016/S0140-6736(16)00215-4

Handelman, S. (1997). *Comrade criminal: Russia's new mafiya*. New Haven: Yale University Press.

Henrich, J., Boyd, R., Bowles, S., Camerer, C., Fehr, E., Gintis, H., et al. (2001). In search of Homo Economicus: Behavioral experiments in 15 small-scale societies. *American Economic Association, 91*, 73–78.

Horne, C., Dodoo, F. N.-A., & Dodoo, N. D. (2013). The shadow of indebtedness: Bridewealth and norms constraining female reproductive autonomy. *American Sociological Review, 78*(3), 503–520. http://doi.org/10.1177/0003122413484923

International Republican Institute. (2004). *Georgian National Voter Study.*

International Republican Institute. (2007). *Georgian National Voter Study.*

Kakachia, K., & O'Shea, L. (2012). Why does police reform appear to have been more successful in Georgia than in Kyrgyzstan or Russia? *The Journal of Power Institutions in Post-Soviet Societies. Pipss.org*, (Issue 13).

Keizer, K., Lindenberg, S., & Steg, L. (2008). The spreading of disorder. *Science, 322*(5908), 1681–1685. http://doi.org/10.1126/science.1161405

Knoch, D., Pascual-Leone, A., Meyer, K., Treyer, V., & Fehr, E. (2006). Diminishing reciprocal fairness by disrupting the right prefrontal cortex. *Science, 314*(5800), 829–832. http://doi.org/10.1126/science.1129156

Krupka, E. L., & Weber, R. A. (2009). The focusing and informational effects of norms on pro-social behavior. *Journal of Economic Psychology, 30*, 307–320.

Kukhianidze, A. (2009). Corruption and organized crime in Georgia before and after the "Rose Revolution." *Central Asian Survey, 28*(2), 215–234. http://doi.org/10.1080/02634930903043709

Kupatadze, A. (2012). Explaining Georgia's anti-corruption drive. *European Security, 21*(1), 16–36. http://doi.org/10.1080/09662839.2012.656597

La Spina, A. (2005). *Mafia, legalità debole e sviluppo del mezzogiorno*. Bologna: Il Mulino.

La Spina, A. (2014). The fight against the Italian Mafia. In L. Paoli (Ed.), *The Oxford handbook of organized crime*. Oxford: Oxford University Press.

Lundgren, R., & Adams, M. K. (2014). *Safe passages: Building on cultural traditions to prevent gender-based violence throughout the life course*. Working Paper, Georgetown University.

Mackie, G. (1996). Ending footbinding and infibulation: A convention account. *American Sociological Review, 61*(6), 999–1017.

Mackie, G. (2000). Female genital cutting the beginning of the end. In B. Shell-Duncan & Y. Hernlund (Eds.), *Female "circumcision" in Africa: Culture, controversy, and change* (pp. 253–282). Boulder, CO: Lynne Rienner.

Meier, S., Pierce, L., & Vaccaro, A. (2013). *Trust and parochialism in a culture of crime*. Technical report, Columbia Business School.

Militello, V., La Spina, A., Frazzica, G., Punzo, V., & Scaglione, A. (2014). D1.1 Quali-quantitative summary of data on extortion rackets in Sicily. GLODERS Consortium.

Mojab, S., & Hassanpour, A. (2003). The politics and culture of "honour killing": The murder of Fadime Sahindal. *Atlantis: Critical Studies in Gender, Culture & Social Justice*, pp. 56–70.

Nardin, L. G., Andrighetto, G., Conte, R., Székely, Á., Anzola, D., Elsenbroich, C., et al. (2016). Simulating protection rackets: A case study of the Sicilian Mafia. *Autonomous Agents and Multi-Agent Systems*, 1–31. http://doi.org/10.1007/s10458-016-9330-z

Opp, K.-D., & Hechter, M. (Eds.). (2005). *Social norms*. New York: Russell Sage.

Orlando, L. (2001). *Fighting the Mafia and renewing Sicilian culture*. San Francisco: Encounter Books.
Ostrom, E. (2000). Collective action and the evolution of social norms. *The Journal of Economic Perspectives, 14*(3), 137–158.
Oxford Dictionary of English. (2012) (3rd ed.). Oxford University Press.
Paoli, L. (2000). *Fratelli di mafia: Cosa Nostra e 'Ndrangheta*. Bologna: Il Mulino.
Paoli, L. (2014). *The Oxford handbook of organized crime*. Oxford: Oxford University Press.
Pezzino, P. (1990). *Una certa reciprocità di favori: Mafia e modernizzazione violenta nella Sicilia postunitaria*. Milano: Franco Angeli.
Pizzini-Gambetta, V. (1999). Gender norms in the Sicilian Mafia 1945–1986. In M. L. Arnot & C. Usborne (Eds.), *Gender and crime in modern Europe*. London: Routledge.
Posner, E. (2002). *Law and social norms*. Cambridge, MA: Harvard University Press.
Punzo, V. (2016). Un approccio analitico al processo estorsivo: Dall'intimidazione alla reazione. In A. La Spina & V. Militello (Eds.), *Dinamiche dell'estorsione e risposte di contrasto tra diritto e società*. Torino: Giappichelli Editore.
Ruff, C. C., Ugazio, G., & Fehr, E. (2013). Changing social norm compliance with noninvasive brain stimulation. *Science, 342*(6157), 482–484. http://doi.org/10.1126/science.1241399
Santino, U. (1995). *La mafia interpretata: Dilemmi, stereotipi, paradigmi*. Messina: Rubbettino.
Scaglione, A. (2008). Il racket delle estorsioni. In A. La Spina (Ed.), *I costi dell'illegalità: Mafia ed estorsioni in Sicilia*. Bologna: Il Mulino.
Schelling, T. C. (1967). The strategy of inflicting costs. In R. N. McKean (Ed.), *NBER* (pp. 105–127).
Schelling, T. C. (1971). What is the business of organized crime. *Journal of Public Law, 20*, 71–84.
Sciarrone, R. (2009). *Mafie vecchie, mafie nuove: Radicamento ed espansione*. Roma: Donzelli.
Shell-Duncan, B., Wander, K., Hernlund, Y., & Moreau, A. (2011). Dynamics of change in the practice of female genital cutting in Senegambia: Testing predictions of social convention theory. *Social Science & Medicine (1982), 73*(8), 1275–1283. http://doi.org/10.1016/j.socscimed.2011.07.022
Slade, G. (2012a). Georgia's war on crime: Creating security in a post-revolutionary context. *European Security, 21*(1), 37–56. http://doi.org/10.1080/09662839.2012.656600
Slade, G. (2012b). No country for made men: The decline of the mafia in post-soviet Georgia. *Law & Society Review, 46*(3), 623–649. http://doi.org/10.1111/j.1540-5893.2012.00508.x
Superti, C. (2009). Addiopizzo: Can a label defeat the Mafia? *The Journal of International Policy Solutions, 11*, 3–11.
Troitzsch, K. G. (2015). Distribution effects of extortion racket systems. In F. Amblard, F. J. Miguel, A. Blanchet, & B. Gaudou (Eds.), *Advances in artificial economics* (pp. 181–193). Berlin: Springer.
Tsitsishvili, D. (2010). *Civil society against corruption*. Georgia.
Tyler, T. R. (2004). Enhancing police legitimacy. *Annals of the American Academy of Political and Social Science, 593*, 84–99.
Tyler, T. R. (2006). Psychological perspectives on legitimacy and legitimation. *Annual Review of Psychology, 57*(1), 375–400. doi:10.1146/annurev.psych.57.102904.19003.
UNICEF. (2008). *Long-term evaluation of the Tostan programme in Senegal: Kolda, Thiès, and Fatick regions* (Working Paper). New York: Section of statistics and monitoring, division of policy and practice.
Vaccaro, A. (2012). To pay or not to pay? Dynamic transparency and the fight against the Mafia's extortionists. *Journal of Business Ethics, 106*(1), 23–35. http://doi.org/10.1007/s10551-011-1050-3
Xenitidou, M., & Edmonds, B. (Eds.). (2014). *The complexity of social norms*. Berlin: Springer.

Chapter 5
Legal Norms Against the Italian Mafia

Vincenzo Militello

5.1 Introduction

Legislation expressly aimed at combating mafia-type crimes in the Italian criminal system has long represented a specific and distinctive set of rules, comprising provisions of criminal law, procedural criminal law and penitentiary law, and including provisions of administrative and civil law of a more preventive nature. However, the law constitutes only one aspect of the complex Italian anti-mafia policy, which indeed results from the combination of several factors, such as human losses, political choices, legislative solutions, evolving case law and reactions of the civil society. In the international arena, among the multiple actors engaged at various levels in the fight against organised crime, the Italian position is characterised not only by the development of specific legal provisions concerning this area, but also by the widespread and active presence in the civil society of citizens, groups and associations committed to encouraging and promoting behaviours opposing the pervasive control of mafia organisations over a territory.

The interaction between different fronts of the anti-mafia action appears to be a necessary approach upon consideration of the internal features of the mafia phenomenon. The actual danger posed by such criminal organisations arises not merely from the use of violence (culminating in the physical elimination of enemies), or in the domination of illegal markets (drugs, arms and waste), but also relates to its ability to infiltrate political and economic environments as well as society as a whole.

Nonetheless, mafia's readiness to make contact with representatives of the civil society and political institutions should not be deemed capable of stopping the overall counteraction against this type of crimes. With reference to the last 30 years, the

V. Militello (✉)
University of Palermo, Palermo, Italy
e-mail: vincenzo.militello@unipa.it

C. Elsenbroich et al. (eds.), *Social Dimensions of Organised Crime*,
Computational Social Sciences, DOI 10.1007/978-3-319-45169-5_5

anti-mafia fight has been strongly established and developed in Italy (albeit through alternating phases often triggered by the reactions to the most serious mafia attacks) both on a legislative and on an institutional level and within civil society. Awareness of the need to promote a culture of legality and anti-mafia behavioural patterns has indeed grown among the citizens.

In this chapter, after considering the initial failure of the Italian criminal code to specifically address the mafia problem (Sect. 5.2) and recalling the first measures of non-criminal nature (Sect. 5.3), the three phases of the anti-mafia strategy shall be summarised. We will then examine the measures originating from the fundamental law of 1982 and enacted in the following decade (Sects. 5.4 and 5.5); the measures which were taken after the murders of judges Falcone and Borsellino in 1992 (Sect. 5.6), with a clear reference to the crucial sector of the fight against illicit proceeds (Sect. 5.7); and those adopted with a view to reorganise the existing copious special legislation produced in this sector, through the enactment of the Anti-mafia Code of 2011 (Sect. 5.8).

5.2 The Long Sleep of the Criminal Code of 1930

As internationally notorious, organised crime and mafia are certainly not a recent issue in Italy. Nonetheless, the Criminal Code of 1930, although quite detailed in its special part, did not contain at first any general or specific provisions to tackle the phenomenon.

Since the mid-nineteenth century, the prevailing opinion in the newly unified Italy considered that the problem should be seen as "one of the many aspects relating to the area of public security".[1] It followed that the earlier anti-mafia measures were mainly developed within a police or military context. As a consequence, the subject matter of mafia was not covered by the Criminal Code—the core of the legal system—and was regarded as an extraordinary problem requiring the adoption of different and ad hoc measures, outside the traditional guarantees provided by criminal law.

This traditional approach can be defined as a normative denial of the specific criminal character of mafia-type organisations, since it considers mafia counteraction as a mere police matter.

The limits of this point of view were soon revealed in terms of both legitimacy and effectiveness. The legitimacy problems related to the protection of fundamental guarantees had emerged since the return of constitutional freedoms in 1948, while the inadequacy of the police approach was increasingly uncovered by the enduring and serious threats posed by the mafia-type organisations in Italy.

[1] For a critical view see Mignosi, *Profili e problemi*, Palermo, 1927, 143. A detailed overview on this former phase is offered by Lupo, *Storia della mafia. Dalle origini ai giorni nostri*, Roma 1993, 19s.

Moreover, Italy has joined the general and international trend characterised by a change in the criminal paradigm, increasingly characterised by the existence of differently structured groups rather than by individual actors who are only occasionally interconnected.

The Criminal Code of 1930 punished the participation in a criminal organisation—characterised by the existence of a group aimed at the commission of various criminal acts—as a distinct offence from the participation of two or more persons in a single offence (complicity). Numerous examples can be found in the wide area of the crimes against the state in the opening section of the Special Part of the Italian Criminal Code. Reference can be made to the offences of subversive association (Article 270), political conspiracy through association (Article 305), and armed gang (Article 306). These articles all criminalise the establishment of a criminal group (or the mere agreement) for subversive purposes. The related problems in terms of the definition of the elements of the offence are only partially unravelled by the reference to the use of the violence or of arms.[2]

The criminalisation of collective entities is more clearly expressed in the area of the crimes against public order, where Article 416 provides for the punishment of three or more people who combine together to carry out criminal offences of any type.[3] The emphasis placed on an open-ended series of offences which can be committed to achieve the organisation's criminal aims is an important expansion of the limited number of specific offences which were previously expressly indicated as the aim of the offence referred to in the corresponding Article 248 of the Criminal Code of 1889.

Therefore, the criminal organisation becomes a model for this category of offences. It constitutes an exception to an important principle enshrined in the General Part of the Criminal Code: namely that the simple agreement (conspiracy) between two or more people to carry out a specific offence is not punishable if the planned offence is not committed or attempted and can be sanctioned only with a security measure (Article 115). This is a principle typical of a liberal criminal system where intentions are not punishable, but which suffers many likewise exceptions in the Special Part of the Code.

---

[2] At this regard see PETTA, *Reati associativi e libertà di associazione*, in *Il delitto politico dalla fine dell'ottocento ai giorni nostri,* Roma, 1984, 198s.; DE FRANCESCO, *I reati di associazione politica*, Milano, 1985; PELISSERO, *Reato politico e flessibilità delle categorie dogmatiche*, Napoli, 2000; MILITELLO, *Offences against the state. Old Basis, Recent Trends and Perspectives on reform.* In Universalis. *Liber Amicorum Cyrille Fijnaut,* T. Spapens et al. (eds.), Intersentia, Antwerpen-Cambridge, 2011, 191s., 198.

[3] Refer to BOSCARELLI, *Associazione per delinquere*, in *Enc. Dir.*, III, 1958, 865s.; PATALANO, *L'associazione per delinquere*, Napoli, 1971; INSOLERA, *L'associazione per delinquere*, Padova, 1983; NEPPI MODONA, *Criminalità organizzata e reati associativi*, in *Beni e tecniche della tutela penale*, Milano, 1987, 107s.; G. DE FRANCESCO, Societas sceleris. *Tecniche repressive delle associazioni criminali*, in *Riv.it.dir.proc.pen.*, 1992, 54s.; DE VERO, *Tutela dell'ordine pubblico e reati associativi*, Milano, 1988; VALIANTE, *L'associazione criminosa*, Milano, 1997; *I reati associativi*, Milano, 1998; *I reati associativi: paradigmi concettuali e materiale probatorio*, Picotti ed al. (cur.), Padova, 2005.

As usual in codified law systems, general rules coexist with wide exceptions: similarly both the offence of criminal organisation—which, as mentioned above, has risen to general model for all associative crimes—and the offence of political conspiracy through agreement (Article 304) both represent an exception to the content of Article 115.

5.3 The First Measures of Non-criminal Nature

The extensive anti-mafia legislation, which nowadays constitutes a specific set of rules within the Italian criminal system, is derived from older measures of a non-criminal nature which had long remained the only regulatory response to this type of crime. The necessity of a criminal intervention in this sector has only gradually emerged in the Italian state based on the 1948 Constitution, confirming the existence of an obvious delay in the political assessment of the seriousness of the mafia phenomenon.

The mafia had returned to manifest itself publicly in Sicily since the landing of the allied troops in 1943, as it is shown by the murders of a series of trade union activists in relation to the issue of land reform (questione agraria), and by the further massacre of policemen (Carabinieri) in Ciaculli (1963), where the use of a car bomb had revealed the organisation's level of violence against public figures and its precursor techniques of terrorist attack.

One of the first signs of political attention devoted to the problem at political level was the creation of the Parliamentary Commission of Inquiry on the mafia phenomenon in Sicily.[4] The work of the commission, enacted in 1948 but only set up in 1962, stretched for over four legislative terms and ended with the presentation of a final report in 1976. The impressive investigative work produced did not result, however, in the drafting of legislative proposals envisaging the creation of specific anti-mafia offences, but rather focused on the strengthening of non-criminal measures.

Ultimately, the raised awareness of the strong presence of the mafia, which had become apparent in the abovementioned massacre of Ciaculli, led to the adoption of a measure which already in its title, "Provisions against mafia" (Law 573 of 31.5.1965), revealed the objective to counter this criminal organisation. The law extended the applicability of preventive measures against a person "suspected of belonging to a mafia-type organisation". These were noncustodial measures restricting personal freedoms (namely special surveillance and the obligation to reside in a designed place) which required lower evidentiary standards than criminal measures and were originally applied within a simplified proceeding of nonjudicial, but administrative nature. Preventive measures are not subject to the traditional guaran-

[4] Similar importance has had, significantly earlier, the Parliamentary Commission of Inquiry in the former Kingdom of Italy (established by law 3.7.1875) which contained great reference to the situation of public security in Sicily, and particularly to the mafia: on this topic refer to the *Relazione della giunta per l'inchiesta delle condizioni della Sicilia*, Roma 1876, especially 111. A private inquiry was, however, conducted in the same period by Franchetti and Sonnino (e.g. Tessitore, *Il nome e la cosa*. Quando la mafia non si chiamava mafia, Milano, 1997, 109s.).

tees provided for by criminal law and constitute a specific system which had developed from the approach originally contained in the law 1423 of 27.12.1956 in relation to persons who pose a threat to public safety. According to this former approach, measures of patrimonial prevention were initially applied alongside the personal measures originally provided for by the law of 1965 (now repealed by the Anti-mafia Code, Article 120, paragraph 1, of legislative decree 6.9.2011 no. 159: see below paragraph 8), but could later be applied independently.

Nonetheless, the introduction of a system of anti-mafia preventive measures did not brush off the existing public feelings of "agnosticism" and "indifference" towards the mafia, as was observed in 1976 by the abovementioned Parliamentary Commission in its final report.[5] After all, the same Parliamentary Commission had been re-established only after the serious attacks in reaction to which the Parliament passed the important law Rognoni-La Torre.

In this period, the failure by the former legislator to specifically address the problem of organised crime[6] reflects the widespread idea (despite the future growing development of supranational legislation in this sector)[7] relating to the impossibility to adopt a clear definition of the phenomenon, which is otherwise regarded as a general umbrella under which several forms of collective criminality and illicit activities exist.[8]

5.4 The Anti-mafia Law of 1982

The continuing and increasing violence exhibited by the mafia since the end of the 1970s and the murders of several politicians, judges and journalists led to a turning point in the fight against the organisation, culminating in the adoption of the anti-mafia law of 1982. The politician La Torre, who originally presented the normative

---

[5] The circumstance that the problem had long been overlooked was raised also by Falcone, *Il fenomeno mafioso: dalla consuetudine secolare all'organizzazione manageriale* (1988), in Id., *Interventi e proposte* (1982–1992), Milano, 1994, 318s.

[6] Refer to Conso, *La criminalità organizzata nel linguaggio del legislatore*, in *Giust. Pen.*, III, 1992, p. 387. Moreover, in Italy the expression was firstly and mainly used in relation to criminal procedure: see Insolera, *La nozione normativa di "criminalità organizzata" e di "mafiosità": il delitto associativo, le fattispecie aggravanti e quelle di rilevanza processuale*, in *Indice penale*, 2001, 20s.

[7] On the issue on a European level please refer to Militello, *Transnational Organised Crime and European Union: Aspects and Problem,* in *Human Rights in European Criminal Law,* S. Ruggeri (eds.), Springer Switzerland, 2015, p. 201s.

[8] The vagueness of the definition represents a long-standing problem: among the many contributions refer to Fiandaca, *Criminalità organizzata e controllo penale*, in Bassiouni et al. (cur.), *Studi in onore di G. Vassalli*, Milano, 1991, 33s.; H.-J. Albrecht, *Organisierte Kriminalität—Theoretischen Erklärungen und empirische Befunde* in Id. et al. (Hrsg.), *Organisierte Kriminalität und Verfassungsstaat*, Heidelberg, 1998, 3s.; Anarte Borrallo, *Conjeturas sobre la criminalidad organizada*, in Ferré Olivé/Anarte Borallo (eds.), *Delinquencia Organizada. Aspectos penales, procesuales y criminologicos*, Huelva, 1999, 20s.; Zaffaroni, *Il crimine organizzato: una categorizzazione fallita*, in Moccia (cur.), *Criminalità organizzata e risposte ordinamentali*, Napoli, 1999, 63s.

proposal, was killed by a mafia bomb attack, but the law was approved only after the further killing of Dalla Chiesa, a General of the Carabinieri, who had been appointed as the Prefect of Palermo precisely to respond to the serious threat posed by the mafia.

This represents one of the most typical traits of the Italian anti-mafia legislation, which is the result of the juxtaposition of special legislative measures enacted, in each case, in response to the most striking mafia attacks. In this regard, the importance of an effective response by the public institutions to the threats posed by criminal organisations is followed by an overall inability to provide a systematic and broad-ranging public intervention against this problem. Moreover, it is also revealed that the mafia phenomenon, far from having extraordinary nature, is however firmly rooted in Italy and also requires the adoption of criminal law instruments.

Nonetheless, this position bears some ambiguity, as to the question whether the continuous emergency posed by new forms of crime is able to justify a departure from traditional criminal guarantees or whether the new crimes should instead be fought within the framework of the traditional criminal system.

The subject animates Italy's criminal policy debate[9] and is also widely discussed on an international level, especially in relation to the specific character of the organised crime provisions within the general criminal system.[10]

The issue is addressed in the law no. 646 of 13.9.1982 (so-called Rognoni-La Torre), which counts at least three objectives: combating mafia as a structured criminal organisation, calling for the immediate confiscation of illicit mafia proceeds and prioritising the fight against the mafia. These objectives are sought through the adoption of offences specifically addressed to the mafia-type organisation; the creation of specific patrimonial preventive measures (seizure and confiscations of illicit proceeds); and finally through the reinstatement of the anti-mafia Parliamentary Commission with the function to monitor the implementation of the new instruments (Articles 32–35).

(a) At the heart of the new system there is the new specific crime of mafia-type organisation (associazione di stampo mafioso) (art. 416 bis) added into the Criminal Code directly after the offence of criminal organisation (associazione per delinquere).[11] The lack of effectiveness of Article 416 in the fight against the

---

[9] A reconstruction of the different positions is offered by Moccia, *La perenne emergenza*, Torino, 1998, specie 88s, and also by Vassalli, *Emergenza criminale e sistema penale* (1995), in *Ultimi scritti*, Milano, 2007, 9s.

[10] On the subject Hassemer, *Kennzeichnen und Krisen des modernen Strafrechts*, in *Zeit. für Rechtspolitik* 1992, 378s. Silva Sanchez, *La expansión del derecho penal* (1998), tr.it. *L'espansione del diritto penale*, Militello (cur.), Milano 2004, 5; Jakobs, *Diritto penale del nemico*, in Donini/Papa (cur.), *Diritto penale del nemico*. Un dibattito internazionale, Milano, 2007, specie 17s.

[11] In addition to the references provided above in Footnote 3 subsequent to the introduction of Article 416 *bis*, also refer to Ingroia, *L'associazione di tipo mafioso*, I Milano, 1993; Turone, *Il delitto di associazione mafiosa*, Milano, 1995; Ronco, *L'art. 416 bis nella sua origine e nella sua attuale portata applicativa*, in B. Romano/Tinebra (cur.), *Il diritto penale della criminalità organizzata*, Milano, 2013, 55s.

mafia had become apparent by the high acquittal rate in the judicial proceedings instituted, in the 1960s and 1970s, against mafia groups, which had the result of strengthening the portrait of the mafia as a new invincible power. Although there are some common elements, such as the minimum number of the members of the association (which must be at least three), the two offences are rather different. Article 416—as all provisions based on the general French model—criminalises the existence of the groups' programme to commit an indeterminate number of crimes, whereas the new anti-mafia offence attaches criminal significance to the organisation's modus operandi (methods) and to the existence of a specific and continuing criminal programme including both illicit and licit activities.

A definition of the mafia method characterising the participation in the mafia-type organisation is thus finally provided, amidst several difficulties encountered by the criminal legislator. Such method consists in the exploitation of the "force of intimidation of the associative bond and of the condition of subjugation and silence (omertà) resulting therefrom" in order to carry out the criminal programme. The focus is on the structure and on the illicit character of the organisation, rather than on the commission of a concrete act of violence or threat aimed at achieving the criminal purposes of the organisation. In this regard it is sufficient that the association has reached a "dominant position" and is able to influence the choices of the citizens, so that in a given social-environmental context they could believe that the members of the criminal groups could to easily commit acts of revenge or retaliation. The mafia method is instrumental in pursuing a programme of various criminal activities, both illicit ("to commit crimes" or more specifically "to prevent or limit the freedom to vote or to get votes for themselves or for others during the elections", subsequently introduced by the Decree Law 1992/36 converted into law 1992/356) and (at least per se) licit (to manage or in any way control economic activities, concessions, authorisations, public contracts and services, or to obtain unlawful profits or advantages for themselves or for any other persons). As for the elements of the offence, while it is sufficient that the organisation is established with a view to pursuing at least one of its criminal purposes, it is not necessary for the proposed aims to be actually achieved. Moreover, the provision provides for the creation of a single offence regardless of the reference to the different purposes to which the organisation is aimed.

The attempt to provide a more precise definition of the criminal conduct is followed by the provision of higher penalties than those provided for the offence of criminal association (also considering the specific aggravating circumstances referred to in paragraph 4.b below); a case for the mandatory confiscation of the goods which constitute the proceeds of the offence is also created, followed by the introduction of the accessory penalty of the automatic withdrawal of authorisations and licences legally obtained for economic activities which can be open to the influence of mafia (law repealed by Article 36, paragraph 2 of Law 19.3.1990 no. 55).

Finally, reference should be made to the explicit equivalence made by the provision between the mafia groups historically rooted in Sicily and all other criminal organisations operating anywhere in Italy, whatever their local titles (camorra, 'ndrangheta, sacra corona unita) (Article 416 bis, paragraph 7): therefore the new offence is not confined to a specific social and cultural context, but attracts a wider definition of criminal organisation, including mafia-type organisations.[12] Besides, the social and cultural references contained in the provisions have been clarified by the continuously evolving case law.[13]

(b) The law of 1982 also introduces a bundle of other specific measures: for example, specific aggravating circumstances which are inherent to the structure of mafia-type organisations: when the organisation is armed or when the economic activities pursued by the organisations are funded, totally or partially, by the proceeds of criminal offences (Article 416 bis paragraphs 4–6).

Other aggravating factors, always related to the activities carried out by the mafia, apply in relation to the offences of aiding and abetting, burglary and extortion when they are committed by the members of a mafia-type group (Article 378, paragraph 2, Article 379, paragraph 2, Article 628, last paragraph, no. 3, Article 629, last paragraph).

The repercussions of the mafia into the legal economy are dealt with in the new offence of unlawful competition with menace or violence (Article 513 bis), aimed at countering the intimidations against competitors operated through traditional mafia style, such as the use of explosive devices or of physical violence and the causing of damages. The scope of such provision is nonetheless significantly narrow, insofar as it covers conducts which are already elsewhere criminalised and the law is thus rarely applied.[14]

(c) The strengthening of preventive measures is the other pillar of the 1982 law. The system of anti-mafia preventive instruments, outlined above (Sect. 5.2), is significantly modified: in particular, the elements of the personal measure relating to the obligation to reside in a certain place or area (obbligo di soggiorno) are more clearly specified, while the number of measures of patrimonial prevention against those who are suspected of belonging to mafia-type organisations is increased. More specifically, the inclusion of the confiscation of goods (Article 14 of the law 646 of 1982) in the body of the former law of 1965 constitutes a radical cultural change. It highlights the importance recognised by the mafia to

[12] See Insolera, *Diritto penale e criminalità organizzata*, Bologna, 1997, 11s.

[13] The legislative definition of mafia method recalls former jurisprudential decisions relating to preventive measures, rather than sociological analysis: e.g. see Cass 8.6.1976, Nocera, in Giust. Pen. 1977. For instance, the vague definition of "omertà" has been intended as "absolute or unconditional refusal to cooperate with public institutions", in a relation of cause and effect with the force of intimidation of the organisation, which is not required to have a necessary general scope, but should only be sufficiently widespread and derive not only from the fear of physical harms, but also from direct and/or symbolic and indirect threats: in practice, it should be a widespread opinion that the (eventual) cooperation with the judicial authority will result in potential retaliation against the person who cooperates and/or her (not just material) goods… "given the structure and operational capacity of the organisation and the existence of other non-identifiable members able to harm those who have opposed resistance (as stated by Cass. Sez. I, 10.7.2007, no. 34974, Brusca, RV237619).

[14] Refer to Palazzo, *La recente legislazione penale*, Padova, III ed. 1985, 236.

capital accumulation rather than other goods such as individual freedom and confirms that members of criminal organisations are more afraid of losing their patrimonies than of receiving a criminal penalty or an administrative ban: an underlying idea which has proven to be right, as is shown by the introduction of further measures which now play a key role in the overall counter-mafia strategy (below Sect. 5.7).

The articulated anti-mafia action envisaged by the law Rognoni-La Torre is followed by a separate measure inspired by the need to develop a coordinate and organic approach in the fight against such a widespread criminal phenomenon. Therefore (by legislative decree 629/1982) a specific administrative liaison body is established, directly appointed by the government, with the aim to ensure the effectiveness of the operations of the public actors involved in the fight against mafia (High Commissioner for the coordination of the fight against mafia crime). The High Commissioner has carried out his activity for almost 8 years (amid acrimonious debate over his direct executive appointment) until its replacement by the anti-mafia investigating office (Direzione Investigativa Antimafia), following a comprehensive reorganisation of all bodies and agencies involved in the sector, proposed by judge Falcone.

Notwithstanding the numerous anti-mafia measures developed in 1982, the risk posed by the mafia in that period was still very high: in July 1983 the judge who had promoted a coordinated approach to anti-mafia investigations and had advocated the necessity to teach students a culture of legality, Rocco Chinnici, was killed in a bomb attack; in the summer of 1985 two anti-mafia police chiefs, Cassarà and Montana, were also murdered.

5.5 Normative Juxtaposition and First Judicial Experiences: The Decade 1982–1992

Following the first judicial decisions on the important new provisions introduced by the anti-mafia law of 1982, Italy returns to focus on the fight against mafia, passing several pieces of legislation, especially of procedural and administrative nature, often in reaction to new dramatic mafia-dominated events. Reference can be made to the following:

(a) Law 29 of 1987 contains provisions regulating pre-trial detention in order to prevent the risk of acquittal of those against whom proceedings had been brought within the so-called maxi-trial, which involved hundreds of defendants who were members of the mafia group of "Cosa Nostra".
(b) Law 327 of 1988 modernises personal preventive measures, modifies the provisions relating to the obligation to reside in a certain place and repeals the diffida (a type of verbal warning).
(c) Law 282 of 1989 (converting the Decree Law 230/1989) regulates the procedure for the administration and the assignment of goods seized and confiscated to the mafia.

(d) Law 355 of 1990 containing "New provisions for the prevention of mafia crime ..." covers various aspects of the relationship between mafia and politics and addresses mafia's infiltrations into the economy and investigations of mafia patrimonies.
(e) The "Urgent provisions on the fight against organised crime, transparency and good governance" (Decree Law no. 324/1990, no. 5/1991 and no. 76/1991 which failed to be converted into law and were all resubmitted; finally Decree Law 13.5.1991 no. 152 converted into law 12.7.1991 no. 203), laying down a range of procedural rules (such as the reinstatement of automatic application of pre-trial detention in relation to organised crime), and prison rules (where inmates convicted of mafia-type offences and other serious crimes are excluded from prison leave while those who co-operate with the judicial authority, even after their conviction, are granted a reduction of penalty).
(f) The critical sector of the protection of those who co-operate with the judicial authorities (pentiti or "Crown witnesses") (who have often played a significant role in numerous anti-mafia proceedings) is the object of specific regulation ("New provisions on kidnapping and on protection of criminals collaborating with justice": Decree Law 15.1.1991 no. 8 converted into law 15.3.1991 no. 82).
(g) Of particular importance is the structural reorganisation of all investigative authorities and the creation of a new super-prosecutor's office. The idea was launched by the unforgettable judge Falcone after his appointment as adviser to the criminal affairs department of the Ministry of Justice. Falcone realised that if organised criminal groups are interconnected, the relevant investigation activity could not be broken down in an indeterminate number of single and unrelated investigations. A National Anti-mafia Directorate, a judicial office composed by a national Prosecutor's Office with the function to coordinate the district prosecutors in relation to mafia proceedings was so created (Law Decree 20.11.1991 no. 327, converted into law 20.2.1992 no. 8). The law also provides for the establishment of the Anti-mafia Investigations Directorate, a department of the Ministry of Home Affairs, composed of members of several police forces (Decree law 29.10.1991 converted into law 30.12.1991 no. 410).

On the judicial front, this period is characterised by the already mentioned maxi-trial against the mafia group of Cosa Nostra, which was instituted in Palermo in 1986. It was the first trial ever with such a high number of defendants (476) and ended in January 1992 in Rome where the Supreme Court upheld the vast majority of the convictions (with more than 80 defendants sentenced to life imprisonment). The symbolic value of the trial was massive: for the first time the mafia had been defeated by the state at the end of a finally organic and coordinate counteraction. The public nature of the hearings put more emphasis on the value of the witness statements of the first pentiti, like those made by Tommaso Buscetta, the mafia leader arrested in Brazil by judge Falcone. Buscetta's declarations had the effect of unveiling, for the first time, the secrecy which had until then surrounded the internal organisation of the mafia and thus represented an important part of the prosecutor's case.[15]

[15] An important direct testimoniancy can be found in the memoirs of the President of the Court of First Instance of Palermo: GIORDANO, *Il maxiprocesso venticinque anni dopo*, Acireale-Roma 2011.

5.6 The Response to the Mafia Attacks of 1992

The convictions served in the maxi-trial (and upheld by the Supreme Court) had certainly delivered serious blows to the mafia; however, the organisation did not surrender and firmly reacted initially by killing a famous Sicilian politician (allegedly punished for his failure to ensure protection to the mafia associates), and subsequently by placing the lethal bombs which would cause the death of the two leading anti-mafia judges and their security policemen: Falcone and Borsellino, between May and July 1992. Along with the strong civil society's reaction which followed the attacks—with marches, demonstrations and citizens' committees[16]—the state responded by passing new legislative measures. More precisely, after the murder of Falcone, the Parliament passed the Decree Law 8.6.92 no. 306 titled "Urgent modifications to the new Code of Criminal Procedure and provisions against mafia-type crime", which was converted into law (Law 7.8.92 no. 356) only after the further bombing killed Borsellino and his policemen.

(a) New provisions were introduced in the Criminal Code with regard to the sensitive issue of the relations between mafia and politics: a wider definition of mafia-type association was provided by the addition of paragraph 3 (Article 416 bis), and the specification of the further aim "to prevent or limit the freedom to vote or to get votes for themselves or for others upon elections". The new and specific offence of political-mafia electoral exchange (Article 416 ter) was also introduced. Unfortunately, the provision sets a high burden of proof on the prosecution and requires the politician to handle the money in exchange for votes, without containing any reference to the possibility that the relevant exchange is made for other economic benefits.[17]

(b) Important changes were also made to the criminal procedure, where a new code had been enacted in 1989 which abandoned the inquisitorial model for a new accusatorial model, providing a stronger system of evidentiary rules and abolishing the figure of the investigating judge (giudice istruttore). The difficulties related to the reconstruction of mafia-related offences in the course of the relevant criminal proceedings led to the abandonment of the traditional criminal procedure rules and to the strengthening of the powers of public prosecutors and judiciary police, as well as those of the National Anti-mafia Directorate and the National Prosecutor's Office; important exceptions were also made to the principle of orality, according to which all evidence should be obtained in the trial hearing and not

[16] The most famous being the committee of the sheets, formed of private citizens who decided to challenge the wall of silence (omertà) and the climate of fear imposed by the mafia, by hanging a white bed sheet on their balconies to publicly express their anti-mafia feeling (for an updated overview see Alaimo, *Un lenzuolo contro la mafia*, Marsala, 2012).

[17] The scope of the offence has long remained limited and the necessity to amend the relevant provision in a way to include the reference to any benefit has only recently resulted in the amendment of the relevant offence by law 17.4.2014 no. 62 (refer to Amarelli, *La riforma del reato di scambio elettorale-politico mafioso,* in *Diritto penale contemporaneo—Riv. Trim.*, 2014/2, 4s.; Squillaci, *Il "nuovo" reato di scambio elettorale politico-mafioso. Pregi e limiti di una riforma necessaria*, in *Archivio penale*, 2014/3).

derived from pre-trial investigations, as well as to the ordinary rules on interception of communications. Provisions increasing the cases of automatic application of pre-trial detention (where serious evidence of guilt exists) in relation to mafia-type offences along with the impossibility to replace such measure with another non custodial measure were also provided. As a consequence, the new law had the effect of affirming the procedural independence of the organised crime proceedings from the traditional criminal proceedings, and developing an autonomous set of rules which would have become the pivot of the Italian anti-mafia system.

(c) A final change relates to the amendment of the so-called hard prison regime (carcere duro), a particularly useful tool in the fight against mafia (Article 41-bis of the law on the penitentiary system). The measure provides for the suspension of the ordinary rules governing prisoner's treatment for those who have been convicted or accused of organised crime and terrorism-related offences. An order under Article 41-bis can be issued by the Ministry of Justice to ensure the protection of public order and the safety of the general public. The treatment—which seemed to be introduced to counterbalance the rehabilitation tendency of the Penitentiary Law 663 of 1986—had originally a temporary and exceptional nature. However, after the massacre of Capaci of 1992, it was extended to include organised crime prisoners and then became permanent (Law 23.12.2002 no. 279). This highly restrictive measure (suppression of almost all contact with the outside world and regime of internal isolation) hinders the implementation of the constitutional principle that penalty should lead to the rehabilitation of the offender, also in relation to mafia prisoners. In this regard, the European Court of Human Rights declared the relating provisions not fully compatible with the European Convention on human rights, even if not in breach of Article 3 which prohibits torture and inhuman degrading treatment.[18]

As is shown by the provisions of hard penitentiary regime, the anti-mafia strategy launched in reaction to the massacres of 1992 developed and strengthened a pattern of already established legislative trends. At the same time, the rush in passing relevant legislation resulted in the commission of serious drafting errors and gaps (as in the abovementioned political mafia electoral exchange).

Important anti-mafia provisions were finally introduced in relation to the protection and assistance offered to the victims of mafia-related crimes. With reference to the crime of extortion, tangible sign of the widespread presence of the mafia in the territory, the increase of the relevant penalties promoted by the anti-mafia law of 1982 (above Sect. 5.4), with evident deterrent purposes, is accompanied by a renewed attention for the position of the victims (as in the offence of usury, which is another type of patrimonial crime characterised by a strong mafia presence).

[18] In particular, as regards the right to respect for private correspondence the court has found a violation of Article 8 HCHR (rule of law principle), as the hard prison regime is applied with an order issued by the Ministry of Justice: see on this European Court on Human Right, judgement of 21/12/2000, Natoli v. Italy Italia; judgment of 28/09/07, Messina v. Italy. On this topic please also refer to Militello, *Der Einfluss der Entscheidungen des Europäischen Gerichtshofes für Menschenrechte auf die italienische Strafrechtsordnung,* in *Strafrecht und Wirtschaftsstrafrecht—Festschrift für K. Tiedemann*, Sieber et al. (Hrsg.), München, 2008, 1421s.

With the idea of improving the effectiveness of the mafia counteraction, the law introduces a new form of state monetary compensation, consisting in the payment of an economic public contribution to the victims of extortion (or usury) and, more general, to the victims of mafia-type crimes who suffer any damages as a result of those crimes (Law 23.2.1999 no. 44: Provisions concerning the Solidarity Fund for victims of extortion and usury; Law 22.12.1999 no. 512 Establishment of Revolving Fund for solidarity with the victims of mafia crimes of mafia type). Nonetheless, the scope of these forms of support to the victims of mafia is limited by a combination of several factors which makes it difficult for the victim to cooperate with public institutions.[19]

Finally, the new criminal policy strategy devised by the legislator also includes amendments to the normative framework governing protection of witnesses and pentiti: in this regard Law 45 of 2001 sets forth more rigid criteria for the admission to the protection programme, reduces the number of offences attracting incentive measures, imposes a deadline of 6 months for the witness to give full testimony (in order to avoid the so-called testimonies in more than one instalment) and extends the protection to informants, namely those who cooperate with the judicial authorities, but who had not committed any crime and often need protection for threat of mafia's retaliation.

5.7 The Key Sector of the Fight Against Illicit Proceeds: Increase of Patrimonial Prevention and Social Reuse of Confiscated Assets

The sector which has been more often reformed over the last year is, once again, that relating to preventive measures. These instruments have gradually assumed some of the characters typical of criminal law, where, for instance, proceedings on an application for the adoption of the preventive measure have become increasingly judicial (e.g. they now take place in public hearings). Nonetheless, they remain essentially different from namely criminal measures and can be classified as having a para-criminal rather than non-criminal nature.

Such specificity has been recently confirmed if one considers that the new anti-mafia provisions have been enacted within the context of the comprehensive reform named "security package".[20] A provision was enacted in 2008 providing for the appli-

[19] In this regard, the data collected in the extortion racket database of the research unit of Palermo in the project "GLODERS" clearly show that the number of cases where the victims resist against the extortion are limited and represent a minority compared to those where victims consent or even cooperate (and are thus complicit) with the authors of the extortive request. These results were found even in relation to cases, object of the research, which occurred in a period where victims already received public economic support.

[20] For an analysis of these trends see Militello, *Sicherheit und Strafrecht: ein "rationaler Dissens". Italienische und transnationale Wandlungen*, in *Innere Sicherheit im europäischen Vergleich. Sicherheitsdenken, Sicherheitskonzepte und Sicherheitsarchitektur im Wandel*, Band 1, Münster 2012, p. 277s.

cation of a patrimonial measure independent of a personal measure and for the confiscation of goods even after the death of the offender (Urgent measures on public security: Decree Law 2008/92 converted into Law 2008/125). This is a clear example of the non-criminal nature of preventive measures, which would otherwise conflict with the principle of personality of penalties, grounded in Article 27 of the Constitution.[21]

The measures at issue have become increasingly important in the fight against illicit patrimonies and constitute the starting point of the overall anti-mafia policy. Nowadays, the anti-mafia strategy is increasingly directed towards the economic assets of the entrepreneurial mafia, under the traditional principle that "crime does not pay". Indeed, the Italian experience reveals the necessity to tackle the reinvestment of illicit financial flows—objective pursued throughout the use of namely criminal measures—as well as the need to target licit capitals, where non-criminal measures are deployed alongside traditional criminal instruments.

As regards the sector of illicit proceeds, we can here recall Italy's money-laundering regulation (Article 648 bis and following), which, since the seventies, has aligned the country with the indications of the main international instruments culminating in the UN Palermo Convention on transnational organised crime of 2000. At the same time, Italy has gradually reformed its traditional measure of confiscation in order to tackle mafia capitals more effectively. Undeniably, the traditional principle of "direct derivation" of the goods or proceeds of a criminal activity shows evident limits in the state's ability to attack the massive illicit wealth accumulated by criminal organisations, in light of the difficulty to prove the distinction between origin and product of the wealth, as expressed by the Latin quote *pecunia non olet*. This brings a paradigm shift where seizure and confiscation should be based on the new principle under which the person is unable to prove the lawful origin of his or her assets or property. Nonetheless, this new approach, which reveals its clear effectiveness in the fight against the illicit proceeds, is problematic insofar as it places an evidential burden upon the defendant.[22]

It is no coincidence that in Italy, despite the wide use of the so-called special (criminal) confiscation, the measures of patrimonial prevention constitute the most effective instrument to counter illicit wealth. These measures, even if applied under judicial supervision, are excluded from the scope of criminal law and are not subject to the same guarantees. As a consequence, paradoxically, the effectiveness of preventive instruments is limited to the Italian territory, in reason of the refusal of foreign judicial authorities to execute abroad seizure and confiscation orders issued by Italian judges. The obstacles to the international judicial cooperation, consequence of the special regime of the Italian preventive

[21] The Constitutional Court has recently considered the question of the legitimacy of confiscation where the subject dies highlighting, also on the basis of the matters referred, the procedural character of the measure and reaffirming the differences between criminal trial and prevention proceedings (judgement 2012/21).

[22] Considerations concerning the existence of similar risks led to Italian Constitutional Court to declare the illegitimacy of an ancillary form of money laundering, namely the unlawful transfer of values, introduced by Article 12*quinquies*, paragraphs 2 and 3 of law decree 1992/306 (judgement of Constitutional Court 1992/48).

measures, represent a major weakness in the fight against illicit proceeds, especially given the increased volatility of international capital flows which can be allocated everywhere in the world and notably moved to the virtual territory of non-cooperative countries.

Nonetheless, the effectiveness of preventive measures should be assessed in conjunction with the specific provisions concerning the social reuse of mafia-confiscated assets, since any intervention in this area could potentially damage the economy and produce negative consequences for innocent victims and, especially, for honest workers. From here follows the important idea that the recovered assets should become part of the circuit of legal economy and be reinvested into the local community for social purposes.

Particularly, the provisions concerning the reinvestment of the mafia-confiscated assets in activities of social nature achieve a dual objective: the restored law enforcement power of the public institutions is followed by the restitution to society of what had been unlawfully taken from it through intimidation and violence. Indeed, this approach offers additional support to the position maintaining that the best results in the fight against the mafia are achieved through the cooperation between institutions and civil society.

5.8 Need for Legislative Reorganisation and Limits of the Anti-mafia Code

An increasing need was felt to reorganise and simplify the extensive corpus of legislation passed over the years in the anti-mafia sector in order to avoid inconsistencies, gaps and contradictions and to assist legal practitioners in the identification of the relevant provisions.

The call for a comprehensive systematic review of the sector is part of a wider debate concerning the enactment of a new criminal code, more respectful of the principles expressed in the Italian Constitution (an objective which is still to be achieved despite the appointment of four ministerial commissions in the past 20 years). In this regard, it was precisely the threat caused by the mafia to be often regarded as the reason for Italy's failure to undertake the codification process embarked in the last four decades by other European countries (from Germany to France, Spain and Portugal).

It is interesting, however, to note how in the sector at issue the necessity for an organic reorganisation of the copious existing provisions in order to limit and avoid relevant inconsistencies, gaps and discrepancies has arisen before and with greater urgency than in other legislative areas. The need to codify the anti-mafia legislation had already been put forward by relevant ministerial commissions, but was finally officially recognised by law 13.8.2010 no. 136 (Extraordinary plan against mafias and Mandate to the Government in the matter of anti-mafia rules), and corresponding implementing decree (Legislative Decree 6.9.2011 no. 159 Code of anti-mafia laws and preventive measures).

Article 1 of the enabling law mandated that the government "should conduct a comprehensive assessment on all existing legislative provisions on organised crime, including the rules contained in the criminal code and in the code of criminal procedure", with a view to achieve the "harmonisation of the considered legislation". The task has not proven easy, given the lack of adequate guidelines and principles, which has caused to delay the interventions on the relevant part of criminal and criminal procedure law concerning the fight against organised crime.

For this reason, the anti-mafia code reorganises only some sectors of the law (and can then be defined as a patchwork-style code). In particular, the new instrument covers the broad and pivotal area of preventive measures and of anti-mafia certificates, establishes the national agency for the management and use of the assets seized and confiscated to the organised crime, while amendments to the Criminal Code, the Code of Criminal Procedure and other anti-mafia laws are only contained in some of its final provisions. The new code introduces some significant changes such as the right of the interested party to request that proceedings relating to preventive measures be held in public, the establishment of rigid, specific time deadlines in proceedings involving seizures, the regulation of the cases where confiscation can be revoked and the clarification of the relationships between preventive and criminal seizure.

The new special code has presented numerous technical issues, which soon led to the start of a new reform path. In particular, the code appears overall vague and incomplete and shows a substantial disregard and misrepresentation of the original objectives set forth by the enabling law, which had called, after all, for the drafting of a codification of anti-mafia laws.[23] It also excludes all provisions concerning specific offences and other numerous procedural provisions.

The logic relationship between the general part of criminal law and its special part is thus upturned as the core of the anti-mafia legislation continues to be the offence of belonging to a mafia-type organisation, envisaged by the Criminal Code in Article 416 bis. This offence has represented a turning point in the Italian strategy against organised crime, attracting the attention of the international community and encouraging the debate on the adoption of a set of supranational obligations to criminalise participation in criminal organisations, which finally culminated in the signature of the UN Palermo Convention of 2000 and were also developed in the context of the European Union's criminal policy.

Thirty years after its introduction, the mafia-type organisation offence remains an essential tool for tackling organised crime: having regard to the Italian prison population—which in the past 5 years numbered approximately 65,000 inmates (well above its rated capacity of 49,000 prisoners)—the overall

[23] See Menditto, *Verso la riforma del d.lgs. n. 159/2011 (cd. codice antimafia) e della confisca allargata*, in www.penalecontemporaneo.it who summarises the main shortcomings of the provisions as follows: "failure to update the proceeding and repeal of provisions incompatible with a 'fair' prevention proceeding, failure to include specific provisions on incompetence, incompleteness of the provisions concerning the administration of goods, disproportionate assimilation to the protection of third parties in the insolvency procedure".

number of mafia prisoners between 2008 and 2009 has never fallen below the 8.6 % of the total and has remained equal to twice the number of those who are detained in relation to the offence of participation in a traditional criminal organisation.[24]

These figures which clearly indicate the overall efficacy of the Italian anti-mafia legislation must be read alongside the number of representatives of the state who died fighting the mafia and of those who still continue their battle to free Italy from the influence of any form of organised crime.

The collaboration between institutions and civil society is undoubtedly the greatest contribution which Italy can offer—even considering the high price paid in terms of human suffering—to the general debate (also at international level) over organised crime.

[24] Prisoners on 31st December classified on the basis of various types of offences in the period 2008–2011:

Type of offence	2008 (and total % at 31/12)	2009 (and % at 31/12)	2010 (and total % at 31/12)	2011(and total % at 31/12)
Mafia-type organisation (416BIS CP)	5.257 (9043)	5.586 (8621)	6.183 (9097)	6.467 (9667)
Public order (416)	2.754	2.975	3.175	3.183
Against patrimony	27.345	30.094	32.225	33.647
Anti-drug legislation	23.505	26.931	28.199	27.459
Prisoners at 31/12	58.127	64.791	67.961	66.897

Source: Official Statistics. Ministry of Justice, Department of Prison Administration.

The figures referred above to the relevant types of offences reflect the exact number of offenders. Offenders who have committed more than one of the above offences are counted within each relevant category. It follows that each category should be considered separately and the figures should not be summed up.

Part III
Extortion Rackets in Society

Chapter 6
Mafia Methods, Extortion Dynamics and Social Responses

Antonio La Spina, Vincenzo Militello, Giovanni Frazzica, Valentina Punzo, and Attilio Scaglione

6.1 Introduction

This chapter is devoted to exploring both the empirical results and the policy proposals produced by the project. It describes a database of 535 extortion cases, which is a large but not necessarily representative sample of the extortion situation in Sicily. Some entrepreneurs (including most of those registered on our database) comply with the mafia's pretenses, because they fear costly retaliation. We can refer to them as "acquiescent victims". Differently, others do not act out of fear, but rather because, on the basis of their culture, beliefs and models of behaviour, they believe that paying protection money is "natural" and "normal" in their social community. We can classify this second category as "tame acquiescent". Both behaviours are morally and civically questionable, but do not constitute a criminal offence or attract a criminal punishment. However, the last category comprises entrepreneurs that, apart from paying what the mobsters request, make use and exploit their relationships with them in order to distort market relationships, scare competitors and obtain unfair benefits. We will return to this category—which can be defined as that of the complicit of entrepreneurs—in the final section.

Anti-racket legislation is an indirect instrument (La Spina, 2008a, 2008b) which—along with the direct repression of mafia affiliates—carried out in Italy through a vast array of tools—is aimed at fighting members of criminal organisations by isolating them and draining away their incomes.

A. La Spina
Political Sciences, Luiss "Guido Carli" University, Rome, Italy

V. Militello • G. Frazzica • V. Punzo (✉) • A. Scaglione
University of Palermo, Palermo, Italy
e-mail: vincenzo.militello@unipa.it; valentinapunzo@libero.it

C. Elsenbroich et al. (eds.), *Social Dimensions of Organised Crime*, Computational Social Sciences, DOI 10.1007/978-3-319-45169-5_6

Some of the members of the GLODERS Palermo unit had been formerly involved in a study sponsored by the Fondazione Rocco Chinnici, which also dealt with the subject of extortion and its spreading within the Sicilian territory (the same research was conducted also in Campania: La Spina, 2016a). However, the present research has been based on a more complete and detailed list of aspects, which were meant to grasp all elements and nuances of the dynamics of extortion. This was important especially in view of the creation of agent-based models which rely on a detailed understanding of the dynamic interactions of agents.

Alongside Sicily our analysis covered also Calabria and revealed the existence of several differences in the extortive conducts carried on within these two regions, which are presented in this chapter. For instance, the 'Ndrangheta rarely accepts to bargain down the amounts of money requested, while Cosa Nostra members are more inclined to negotiate also as a way to gain some trust from entrepreneurs. Therefore, apparently acquiescent victims are more numerous in Calabria, because the 'Ndrangheta tends to avoid dialogue and prefers sheer imposition. On the contrary, Sicilians probably tend to engage in discussions and are inclined to peaceful cohabitation, which reflects the attitude of so-called tame acquiescent entrepreneurs.

The first section of this chapter illustrates the legislation relating to the crime of extortion, as well as the policy concerning the compensation of the damages suffered by entrepreneurs who have reported the racketeers to the police, which is specifically aimed at encouraging as many victims as possible to do so. The following sections are devoted to Sicily and Calabria, respectively. The last one presents some social and legal responses to the phenomenon.

6.2 Juridical Aspects on Extortion

Although extortion is a classic crime against property, it has been the subject of growing interest in the recent Italian legislation, as part of a wider attention on the fight against organised crime of mafia type. As has been recently shown through the cases collected by the Palermo research within GLODERS and in line with other previous studies conducted in specific Italian regions (Di Gennaro & La Spina, 2010; La Spina, 2008a), extortion is a typical field of mafia-type illegal activities. Its distinctive feature consists in combining the structural traits of the mafia organisation—the creation in a specific territorial and social context of a widespread climate of intimidation—with the elements of the crime of extortion, where the victim is forced to perform a specific action under threat of a harm. The gradual understanding of this criminological phenomenon has placed extortion among the sectors of the Italian criminal policy against the mafia. The numerous reforms which have interested the offence of extortion have not altered the description of the relevant criminal behaviour, which has remained the same since it was first defined in the Code of 1930: the use of violence or threat, to compel someone to do or omit to do any act in order to procure for themselves or others an unjust profit and to cause other damages. Differently, the sanctions envisaged for this crime have been the subject of many changes: first of all the minimum penalty has been increased from 3 to 5 years; if there are special aggravating circumstances the minimum penalty is increased from 4 to 6 years and the maximum from 15 to 20 years. Equally, the related pecuniary penalties have also risen in relation to both the main offence and its related aggravating factors, such as the circumstance (introduced by law 1982/646 on mafia-type association), where the violence or threat is perpetrated by a member of such association. Thus, today the most typical form of this crime—namely extortion committed by members of mafia associations—is punishable by a penalty including imprisonment up to 20 years and a fine of euros 15,000. However this traditional way to fight criminal phenomena—strategy of deterring crimes through raising penalties—affects only the choices of potential criminals.

In the structure of extortion an important role is also played by the conduct of the victim. The victim of this offence is the person who, although under the threat of violence, carries out a conduct which can be harmful to himself or herself or others. It should be considered that the condition of constraint resulting from the use of violence or threat against the victim must still leave him or her with a minimum of autonomy in the choice of his or her own behaviour. In the absence of such a requirement, the offence committed will be a robbery (rather than an extortion), because the offender can directly obtain an unfair advantage. After recognising the strong link between extortion and mafia, the Italian penal system has therefore turned its attention to the position of the victims of this specific criminal activity (which is similar to that of the victims of loan sharking, another crime strongly related to mafia-type organisations). For this reason, since 1999 the Italian State has provided a special financial compensation for the victims of extortion, consisting in the reimbursement of the damages arising from extortion activities (the same measure is provided for usury victims: so-called Fondo di solidarietà: Law 44/1999; and a similar fund was established also for the victims of mafia-type crimes: law No. 512/1999). However, the introduction of these important forms of support has not had a great impact on the attitudes of the victims of extortion with regard to the lack of reporting and the consequent difficulties to identify the authors of the relevant crimes. In this respect, the cases collected from the database on extortions produced by the GLODERS research clearly show that the resistance of the victim is limited. These cases represent a large minority compared to those where there was acquiescence or even complicity with the authors of extortion. These dynamics can be observed even in relation to the relatively recent cases which are now included in the database and that were thus committed at a time when the existence of public financial support to the victims was widespread. The lack of cooperation with the law enforcement agencies has produced a significant reaction from civil society, at least in some local contexts (such as Palermo). In particular, citizens' associations launched and maintained for several years a campaign aimed at raising public awareness indirectly avoiding those businesses which suffered extortion, without reporting the crime to the police.

This implicitly means that the measures in support of the victims of extortion-related activities (the already mentioned "Fondo di solidarietà") can be considered as an entitlement to request the state's compensation for the damages suffered. It derives, as a consequence, the existence of an obligation in this respect, and the punishment, on an economic level, of those who are affected by the crime, but do not collaborate with the state to tackle this urgent problem.

The transformation of the perception of the subject of an extortive request from victim into an accomplice of the criminals has also been reflected in legislation, followed by the exclusion from professional associations of employers of those members who do not cooperate with the authorities. As a result, the victim's failure to report extortion has been considered illegal and directly punishable (although not under criminal law). The reason underlying this choice reflects the view that this widespread attitude among the victims represents an obstacle to the contrast of the phenomenon and constitutes a contributing factor to the limited effectiveness of the relevant sanctions which are indeed only applicable to fully ascertained facts. Employers who do not report the crimes of extortion and of aggravated extortion committed against them by third parties can therefore be excluded from bidding for public contracts (as provided by art. 2, paragraph 19, of law July 15, 2009, No. 94, which amended Article 38 of the Public Contracts Code) (Militello & Siracusa, 2010).

Despite the encouragement of victims with the authorities appears crucial in order the contrast the phenomenon more effectively, the compatibility of penalties imposed on the victim of extortion with a rational criminal policy is questionable. The most problematic point is the intrinsic contradiction between the position of the victim who is subject to an extortion request, in a context characterised by the presence of the mafia—which structurally implies a climate of intimidation and silence—and the threat of punishment by the state in order to induce the same subject to act as he or she was entirely free to choose and in control of his or her actions (as he could be completely removed from his or her context). It is not coincidence that the abovementioned penalty has remained largely unapplied. Thus, the question remains open so far as to the identification of more adequate legislative forms of support which could promote, more effectively, through positive actions, the victims' cooperation with the criminal justice system.

6.3 Sicily and Calabria Extortion Database

The empirical investigation of extortion racket systems (ERSs) raises some methodological issues. Firstly, a study of extortion, as well as the analysis of its spreading with reference to different forms of extortion, involves the collection of large amounts of data and information about the assets of criminal organisations on the one hand and on the socio-economic and cultural aspects of territorial contexts on the other.

With regard to the quantification of extortion-related crimes, one of the major problems relates to the existence of the "high dark" number, where only a few cases of extortion are reported by the victims and discovered by the police forces.

In this scenario, some methodological questions are associated to the empirical investigation on ERSs.

Up to now, most of the research on ERSs has focused on measuring the *quantum* of extortions (Di Gennaro & La Spina, 2010; La Spina, 2008a, 2008b). In this perspective, several surveys have tried to quantify the amount of money and goods extorted in Italy (e.g. surveys conducted by CENSIS; SOS Impresa, 2007), by means of victimisation studies. Thousands of entrepreneurs have been interviewed in order to address their perception about the spread of extortion in a specific territorial area.

This kind of methodological approach has been criticised for being unable to measure the real impact of extortion as the data obtained cannot be considered reliable (La Spina, 2008a, 2008b). Recently, empirical research on extortion tried to overcome some of the limitations of the current studies employing other empirical sources, different from the self-report data, to measure the impact of ERS (Di Gennaro & La Spina, 2010; La Spina, 2008a, 2008b; Transcrime, 2013). In this regard, the analysis of judicial documents, containing official results of judicial investigations (conducted through wiretaps, testimonies of the collaborators of justice, patrimonial surveys, analysis of documents owned by Mafiosi, etc.), seemed to be a more objective source of information about extortions than the victimisation surveys and to offer a more realistic picture of the phenomenon.

In this methodological approach and moving on from the previous empirical researches conducted by the Rocco Chinnici Foundation in 2008 and 2010 (Di Gennaro & La Spina, 2010; La Spina, 2008a, 2008b), the specific objective of the empirical analysis on case studies on extortions was to deepen the quality of the knowledge on the ERSs in territorial areas in which mafia-type organisations are embedded. We have therefore selected two study cases in Italy and investigated two regions, Sicily and Calabria, traditionally characterised by the presence of mafia-type organisations.

Afterward, the collection of empirical sources has been crucial in our research in order to enrich the quality of empirical evidence and to expand our understanding of extortion. We believe that it is not sufficient to determine the *quantum* of extortion, and have thus decided to focus our study on the dynamics of the process itself. Our main research questions were the following: how the extortion process takes place; what type of actors are involved in the process of extortion; what are the factors that intervene into the process and under what conditions are extortions exerted; and what are the most significant territorial differences in the management of extortion.

Our investigation aimed at answering the above questions and at shedding more light on a hidden phenomenon such as that of extortion racket.

The empirical research followed two different directions: on the one hand we have conducted some in-depth interviews with entrepreneurs who had reported the crime of extortion to the police and with judges involved in anti-mafia investigations. On the other hand we have considered the findings of judicial sources. Particularly, in relation to the latter, we have collected judicial, police and court documents related to the most important anti-racketeering operations conducted in Sicily and in Calabria against mafia-type organisation between 2007 and 2013. The analysis of the abovementioned judicial documents has been conducted in order to gather information about reported cases of extortion and to elaborate a database on extortions.

Overall, we have analyzed a total of 631 cases of extortion (actual or attempted) in Sicily and Calabria. Significant differences exist between the data that we have analysed and classified and the empirical data on extortion previously collected in other studies (La Spina, 2008a, 2008b, 2010). In particular, the present database covers the past 6 years. Also, it contains more recent and complete data and takes into account a larger number of people and cases.

The Sicily and Calabria extortion database represents a relevant achievement in the field of the research on extortion racket and is one of the main results of the GLODERS project.[1] Compared to the data on the phenomenon of extortion offered by previous empirical researches in these territorial areas, the database, based on 631 cases of extortions, allows us to gather a great amount of information about extortions in Sicily and Calabria.

As for the Sicilian section of the database, in contrast to existing data produced as a result of the research carried out by the Fondazione Rocco Chinnici in 2008 which was aimed at quantifying the costs of extortion-related crimes in Sicily by calculating the amount of money and goods extorted (La Spina, 2008a, 2008b), the current updated and enriched database, with its 535 extortion cases, includes much more information regarding the extortion process. Special attention has been devoted, for the first time, to the analysis of criminal groups, their internal networks and relationships and their detailed concrete behaviours in the management of extortion. With regard to the Sicily database, the data relating to Palermo have been the first to be collected. More specifically, we have collected and analysed a total of 144 cases of extortion (actual or attempted) registered in the District of Palermo. In relation to the analysis of the data from Palermo, a case study analysis on the mafia-type groups (or families) involved in extortions has been conducted in order to augment the knowledge of the structure of criminal organisations.[2]

As for the data on extortions in Calabria, this is the first database ever developed in this territorial area. The database constitutes an important source of information in order to offer a more comprehensive and broad view of the phenomenon in Calabria.

As for the methodological issues encountered in the construction of the database, considering that extortion is a crime characterised by a high dark figure, we could not rely on a random sampling. Therefore we have opted for a *purposive sampling* following the criterion of a purposive territorial coverage.

[1] The database is available as an SPSS file at the repository "Datorium" managed by the GESIS Data Archive for the Social Science and can be downloaded at the following link: https://datorium.gesis.org/xmlui/handle/10.7802/1116.

[2] For the case study analysis on the mafia-type groups see Scaglione A. (2014) "Understanding mafia-type organisations: resources, dimensions and leadership styles", in V. Militello, A. La Spina, G. Frazzica, V. Punzo, A. Scaglione (2014), *D1.1 Quali-quantitative summary of data on extortion rackets in Sicily*. GLODERSGLODERS report; Scaglione (2015), in A. La Spina, G. Frazzica, V. Punzo, A. Scaglione (2015), *How mafia works*. The analysis focused on ten case studies of mafia-type organisations (MTOs) emerged from the database: six of which occurred in the city of Palermo, and the remaining four in the province (three cases relate to the province of Palermo, one to the province of Trapani).

As for the Sicily data, we aimed to represent the proportional distribution of the cases between eastern and western Sicily. The only two provinces not represented in the database are Caltanissetta and Enna (see Fig. 6.1).

As for the Calabria database, instead, we focused on the territorial area of the chief town of the Region, named Reggio Calabria, which is the district of Calabria characterised by the highest rate of mafia rootedness.

The database consists of a column chart where each row describes one extortion case. For privacy reasons, we ensured that all sensitive data (first of all the names of the people involved or who could be identified also indirectly through the localisation of the events) were adequately protected. For this reason the names of extortees have been replaced by pseudonyms. The first columns contain some technical facts (unique case number, reference to the original source, region, case number within the regions Sicily and Calabria), which is followed by information about the time when the case occurred, the pseudonym of the extorter, his or her (all extorters are male) role in the organisation and the name and territory of the mafia family—or the so-called mandamento he or she belongs to. These data are followed by information about the victims, and the category of business they deal with. The type of business is coded in accordance to the official Italian coding scheme (AtEco). The final group of data related to the place where the extortion has taken place.

After further technical information about the extortion cases, the factual events of the case are described. Most variables come in two forms, both the original textual description of what happened and how it happened and a recoded variable which is more suitable for quantitative analyses. For this reason the database

Fig. 6.1 Distribution of the extortion cases in Sicily

includes quali-quantitative data in order to offer not only a quantitative, but also a qualitative representation of the phenomenon. More specifically, the quantitative data related to the quantum of extortion were combined with qualitative information on the process of extortion, the type of actors involved and the dynamics by which it arises: types of intimidations, presence of negotiation, types of mediator, etc.

In particular, the features described in these variables encompass:

- Whether the extortion was only attempted (and unsuccessful from the point of view of the extorter) or completed, i.e. the victim actually paid
- Whether the request was for a periodic or a one-off payment or both and what the amount was (the amounts of periodic and one-off payments are not always comparable as some were only defined in terms of percentages of the victim's income or in terms of obligations the victim accepted to assume, i.e. to employ a relative of the extorter, etc.)
- Whether an act of intimidation occurred and whether it was directed to a person or to a property
- Whether the extortion request was brought forward by direct personal contact or by any form of indirect communication
- Whether any and/or what type of negotiation occurred between extorter and victim, and whether a mediator was involved
- How the victim reacted: was he/she acquiescent, resistant or conniving
- How the law enforcement agencies became involved in the case (own observation, denunciation, etc.)
- Whether the extorter was arrested, charged, detained or sentenced (these data contain a high percentage of missing information, partly due to the fact that some investigations or proceedings are still ongoing or as a consequence of incomplete documents)

6.4 The Phenomenology of Extortion Rackets in Sicily: Territorial Differences and Main Empirical Results

The database on Sicily includes a total of 535 cases of extortion. It collects cases from almost all the Sicilian provinces. 51 % of cases occurred in eastern Sicily, while 49 % of the episodes surveyed were established in western Sicily. The data collected are related to the major cities of the island: Palermo, Catania, Messina, Trapani, Agrigento and Siracusa. In particular, the database contains several sources of information: on the crime, on the offender, on the territory, on the dynamics of the crime and on the methods of extortion. Therefore, the database is a valuable tool which can hugely increase the knowledge on the persistence and spread of mafia-type organisations. Territorial distinction is relevant because, contrary to what is observed in relation to eastern Sicily, not all MTOs therein operating are linked to *Cosa Nostra*, many of them acting independently. This is also the reason, for instance, for the existence of a greater level of conflicts in the borough of Catania,

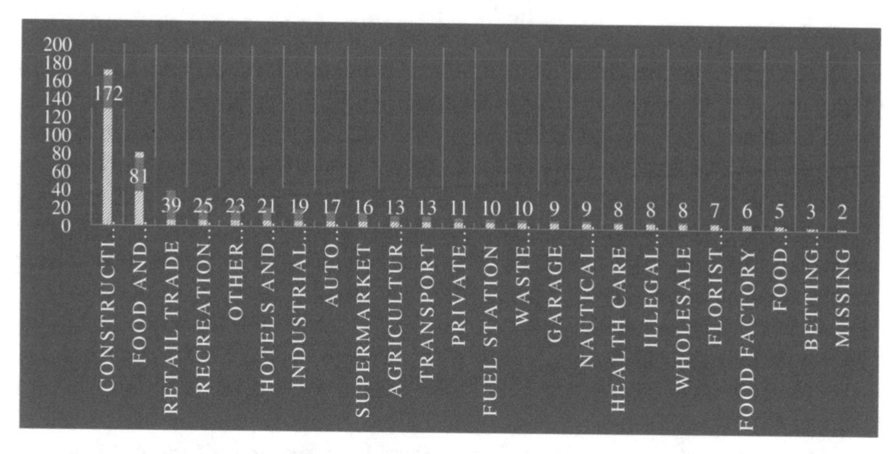

Fig. 6.2 Extortion cases by economic activity

the main city of this area, where cases of extortion were classified by the type of business, mostly relating to building activities and correlated services, food and beverages and retail trades (Fig. 6.2).

More than double the extortion cases collected were consumed (382), while less than a third (153) were actual. MTOs therefore have a high rate of success (at least two of the three attempts of extortion are successful). Some interesting differences emerge from the territorial analysis. The number of extortion consumed is higher in eastern Sicily than in the west. This finding could be related to limitation of our database, but could also suggest a higher incidence of the anti-mafia movement (*Addiopizzo*) in western Sicily.

Intimidation is at the basis of the mafia method. Intimidation is widespread (521 cases out of 535). Almost every case of extortion (89 %) contains the element of intimidation. As is known, the mafia families rely on and apply a power of intimidation to extort periodic payments from legitimate business enterprises. However, in order to better understand how the process of extortion takes place, it is useful to analyse the different types of intimidation.

Firstly, intimidation can be *implicit* or *explicit*. In most cases, intimidation does not involve any use of violence or damages against people or property. Here, intimidation is *intrinsic*, and is implicit in the mere presence of a Mafioso. It relies on the reputation of the mafia men and on the mafia itself. The mere presence of a mafia man, representative of the MTO, constitutes a form of intimidation which is *intrinsic* to the structure of the MTO which entails the expectation to use violence. Consistently with a widespread implicit intimidation, data show that violence is not used in 64 % of cases, while a verbal threat characterises 36 % of cases of explicit intimidation,

The combination of this empirical evidence can assist in determining the level of reputation held by Cosa Nostra in the district of Palermo. In fact, "reputation is earned by delivering promised protection, which implies keeping promises using information effectively and resorting to violence when necessary" (Matsueda, 2013,

p. 312). Mafia considers reputation as an asset that exempts the "firm" from having to prove quality and reliability in every transaction and also shields it against other competitors. Moreover, according to Gambetta (1993), reputation also reduces the so-called costs of production of the mafia, where the stronger the reputation the less likely it would be that the mafia will have to use up resources, such as violence, to guarantee protection and to maintain a high reputation and the control of the territory. Thus, the low percentage of explicit intimidations indicated the level of reputation of the mafia. Dasgupta 2000 suggested to measure reputation as a dichotomous rather than as a continuous variable. Therefore, it is impossible to remedy the loss of reputation of mafia men, and that explains why mafia men exert punishments to endangering reputation with drastic executions often arranged with horrific and theatrical detail to discourage future opponents (see Chinnici & Santino, 1989 in Gambetta 1993, p. 46).

In order to deeply understand the connection between types of intimidation carried on by mafiosi and their reputation, a distinction between *intimidation against people* and *intimidation against property* needs to be made. In this perspective, data suggest that intimidation is almost equally directed both against people and against property. Only a small number of cases (6 % out of the total cases) show the presence of both types of intimidation. This evidence could be misleading outside of a consideration of different forms of intimidation against people. Most intimidations take the form of warnings, while damages are inflicted only in 21 % of cases. However, rather than being an alternative, the type of intimidation is generally related to the degree of resistance of the entrepreneur: if a warning is not enough, the infliction of damages will almost always follow. On the basis of the distinction between intimidation as a warning and intimidation as a proper damage (both directed against people and against property), we will now analyse the types of intimidation against people. In this regard, only a small number of cases show the existence of a proper *damage* against people, where in most cases, intimidation against people takes the form of *warnings*. The presence of intimidation against people takes generally the form of a verbal threat that is a symbolic warning which normally characterises the first approach between the mafioso and the entrepreneur. The value of the warnings in these cases relies on the reputation of the mafia-type organisation which controls the territorial area. The higher the *embeddedness* (and also the credibility) of the mafia-type organisation over the territory, specifically the mafia family which controls that territory, the higher is the symbolic value of the warnings given by the extorter. In fact, as pointed out by Matsueda, although periodic demonstrations of violence would reinforce the value of the mafia's reputation, "even in the absence of such demonstrations, reputations persist because customers are unlike to challenge them on account of the high costs of violence" (Matsueda, 2013, p. 312). As a consequence, the request of "pizzo" acquires the form of a protection racket (Gambetta, 1993; Santino, 2006; Sciarrone, 2009), where the mechanism of protection extortion is based on the reputation of the mafioso.

Thus, intimidation against people can be viewed as an escalation of acts of intimidation which start from symbolic warnings (mainly verbal threats) and evolves, in a small number of cases, into the infliction of damages such as physical aggression.

The distribution of frequency in the case of intimidation against people in fact shows an asymmetric shape in which the largest number of cases thicken around the verbal threat and only a relative limited number of cases show a physical aggression. The latter case is almost always associated to cases of resistant behaviour against extortion by the concerned entrepreneurs. Moreover, in the cases of resistance against extortive request, intimidation against people is often accompanied by some form of intimidation against property, mainly arson. In fact, contrary to the intimidation against people, intimidation against property implies in most cases the infliction of a damage. Only in a relative small number of cases (37 %) extorters just use warnings in relation to properties.

Thus, the most common form of warning against property is represented by the use of glue which is placed into the locks (46 % of cases) of the businesses' premise, an act which always holds a symbolic meaning. Among the "symbolic damages", a common method is "to deploy young people to place the quick-setting glue type 'attak' in the locks of the shop, the fuel tanks near the firm or gun shells in an envelope, and so on" (Scaglione, 2008, p. 89). The placing of glue in the locks, which can be seen as a mafia signature, has a high symbolic value confirming the presence of the mafia into the territory. In fact, one of the first acts perpetrated by the mafia against an entrepreneur who refuses to pay *pizzo* is "to fill the shop's bolts and locks with glue during the night. The next morning, the shop manager has to call for a carpenter or the police to get access to the premises, which results in loss of business" (Vaccaro, 2012, p. 28). Only in a relative few cases warnings are a proper sabotage. Among damages against property, data show a prevalence of arson. In most cases arson is perpetrated against cars and only in few cases against shops or houses. Moreover, as for physical aggression in the case of intimidation against people, arson is always associated with a resistant behavior by the victim of extortion. Generally, negotiation is infrequent and on this topic we need to further investigate our data as negotiation requires a deep analysis of the relational dynamics.

As far as the negotiation phase is concerned, a bargaining between extorters and extortees takes place only in a small number of cases: and in particular only 77 out of 535 cases show some sort of negotiation. Bargaining is then absent in almost all cases (458 out of 535 cases show no negotiation). In the few cases in which negotiation occurs, the bargaining between extorters and extortees deals with the reduction of the initial request or—in some cases also with an extension of the time for the payment of the pizzo. In the first case, the bargaining often results in the reduction of the initial request up to 50 %. Furthermore, negotiation is almost always unmediated. Only in a few cases the negotiation is managed with the presence of a third subject, who plays the role of mediator. Negotiations between extorters and extortees are then generally direct. Moreover, as for the presence of a third person who mediates the relation between extorters and extortees, only in few cases interaction is mediated by a third subject. In those limited cases the mediator is often a fellow countryman. Sometimes the mediator is also chosen by the victim. It is useful to point out that the presence of a mediator does not mean that some bargaining will definitely take place, given that in most cases, even if there is a mediator, there is no negotiation.

The extortion racket is the typical activity of the MTO, as it is the most suitable tool for the control of the territory. Indeed, as pointed out by the literature, there can be extortion without mafia, but there is no (nor there could be) mafia without extortion. According to Monzini "the specificity of extortion is that it continuously penetrates into areas of existing markets, licit or illicit, influencing their operation". Thus, extortion is the criminal activity "bounded to a specific organisational context, which overlaps and which tends to become an integral part: *parasitic* or *symbiotic* in its forms, is always linked to the local level" (1993, p. 1). Different types of extortion can be found in literature, corresponding to different typical methods of request. Firstly, the demand can be *predatory* or *protective* (Monzini, 1993; Scaglione, 2008; Sciarrone, 2009) on the basis of the strategy implemented by the mafiosi, on the relationship that the MTO establishes with victims/recipients of the extortion request and on the type of structure of the organisation (for a review see also Punzo, 2013). *Predatory extortion* is the simplest form of extortion, which is episodic in nature. Mostly, it is observed in areas characterised by the presence of criminal organisations in disarray and it is generally addressed to wealthy, or also vulnerable, entrepreneurs (Scaglione, 2008). The threat of actions against the property or individuals is followed by a monetary request, often disproportionate, where "the extortee has no interest in appearing and does not assume, therefore, any responsibility to protect entrepreneurs" (Monzini, 1993, p. 15). The threat, therefore, is almost always anonymous, commenced with caution and sometimes improvised. In this case the consumption of extortion rarely involves the payment of the amount demanded, and often ends up with the payment of a minimum amount in goods or in cash (one time), which is generally considerably lower than the initial request.

Moreover, an extortion request can be made to all entrepreneurs who operate in a specific territorial area subjected to the mafia control or, otherwise, only to some selected entrepreneurs. In the case of the *protective demand*, the aim of the MTO is to maintain the control over the territory through *systematic extortions*. In this perspective, "it will appear rational to the mafiosi to impose the correspondence of a periodic payment by each extorted, mostly agreed on the basis of the economic possibilities of the victims" (Scaglione, 2008, p. 95). In turn "the organisation is committed in providing protection from the risk of theft, damage, theft etc." (Scaglione, 2008). The perception of threat, in the case of systematic extortion by the mafia, is very high, such as to place the recipients of the request in a position of subordination or acquiescence. In such cases, the request "tends to maintain a threshold of coexistence and tolerance between criminal organisations and their environment of reference" (Monzini, 1993, p. 17). Among the *protective extortion*, which typically characterises the MTO in a given traditional area, different forms and types of *demand/consumption* have been distinguished (Monzini, 1993; Scaglione, 2008; Sciarrone, 2009).

The demand can be an *episodic* or a *periodic* request of money or, in some cases, of goods or services.

Data show that periodic request is more widespread than the episodic one. As we know, the request may be single or repeated over time. The "*una tantum*" request is equivalent to the so-called *messa a posto* "put in the right place", namely a kind of

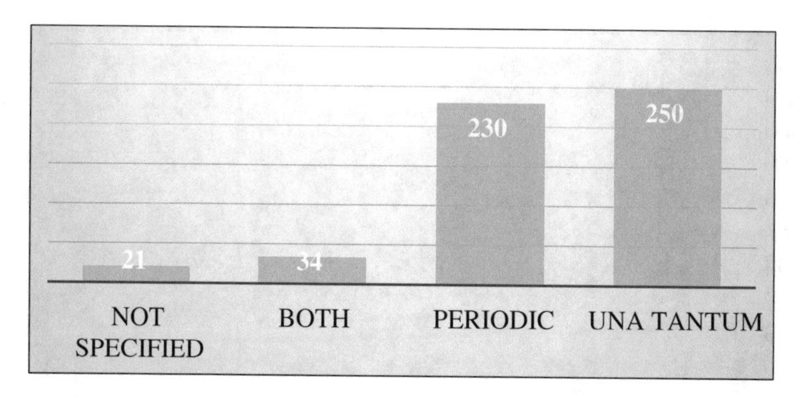

Fig. 6.3 Episodic and periodic demand

amnesty for past missed payments or for opening a new business. In general, the periodic payment occurs during the holidays (Christmas and Easter) or can also follow a yearly, monthly and even daily or weekly trend as observed in some of the cases analysed (Fig. 6.3).

The most significant differences are recorded with respect to the economic sector. In the secondary sector, which also includes construction activities, the request is generally one-off, while in the tertiary sector it takes the form of periodicity. Periodic and episodic demands can be made in the form of goods or services, or also in the form of certain conditions imposed on the business management, such as preferred suppliers and personnel to be hired. As regards the type of request, seven times out of ten it is monetary. The quantum may vary from few hundreds to many thousands of euros. However, as we have already noted, there are other forms of payment: imposition of services, payment in goods, staff recruitment, forms of restriction of the economic freedom and attempts, in some cases succeeded, of expropriation of the company. In the "multiple" category we have grouped all those cases where the request takes different forms (e.g. staff recruitment and money, or money and goods, etc.). We distinguished the *parasitic* demand, in which extorters get something from the extortees (money or goods) from *active demands*, when extorters impose something to the extortees, such as some personal performance: supplies, purchase of products and services, purchase of video poker machines, recruitment of employees, etc. 72 out of 100 cases entail a parasitic demand of money. Among active demand (25 %) imposition of supplies, purchase of products and recruitment of employees are almost equally present. Most entrepreneurs pay and do not report. Among resistant entrepreneurs just over more than half have reported the extortion to the police. Acquiescence behaviour is slightly more widespread in eastern Sicily. Resistant entrepreneurs are in greater number in western Sicily. This could be related, as already said, to the greater impact of Addiopizzo and similar associations. With respect to the economic sector, we observed that in general the acquiescent behaviour is more frequent in the secondary sector. In any case, these are marginal differences that we should consider with caution on the basis of the available data (Fig. 6.4).

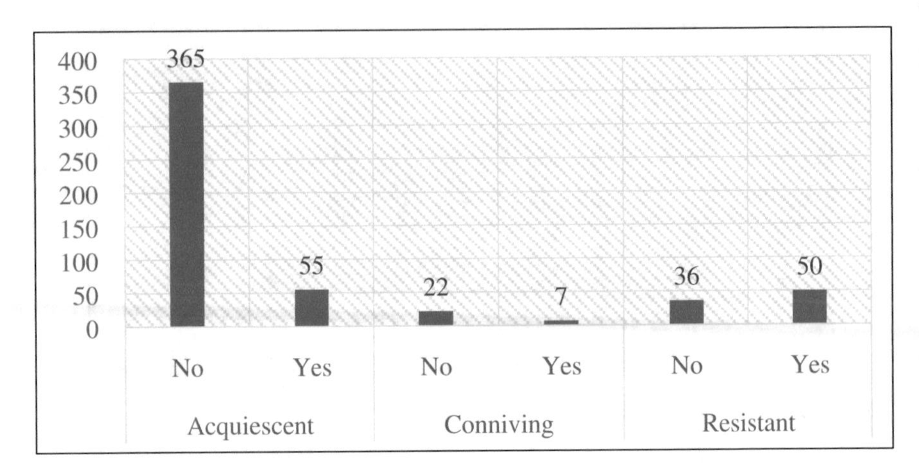

Fig. 6.4 Reporting behaviour by victim type

6.5 Extortion in Calabria: Some General Information About 'Ndrangheta and Principal Empirical Results

Even in Calabria, as well as in other areas where the research has been conducted, the phenomenon of extortion racket is widespread. As regards the data collected in Calabria, we can count 94 cases of extortion. We have also collected information from interviews carried out with judges who work for the court of Reggio Calabria. Prosecutors have also provided judicial documents which have represented the start of the database containing the cases of extortion in Calabria. The way in which the 'Ndrangheta imposes costs on entrepreneurs is frequently shown under various forms not very different from those that we have found in relation to Sicily. In many cases, however, intimidation is very well hidden behind relational dynamics that are usually very hardly to observe. These types of relationships are strongly connected with the local cultural background. Obviously, we do not intend to refer only to the cultural elements, but we certainly have to recall the strong symbolic meaning of extortion in Calabria, also in terms of control of territory. Among the main consequences suffered by entrepreneurs, it should be noted that very often those who do not pay are isolated and cut off from the market. Thus, the decision to report, in many cases, is also related to the value of legality and the desire of change. However, even if the scenario is constantly changing, the main successes in this area are the investigative results achieved by the law-enforcement authority.

An interview has also been conducted with a member of the Observatory on the 'Ndrangheta and with two entrepreneurs who reported the crime to the police. These interviews have helped us to better understand data collected from judicial document that we have analysed in order to build the database on Calabria. The interviews were analysed according to a qualitative approach. We have selected the most significant portions of the text with reference to the various topics covered by the questions. We carried out in-depth interviews and the answers of the entrepreneurs

Fig. 6.5 Type of extortion

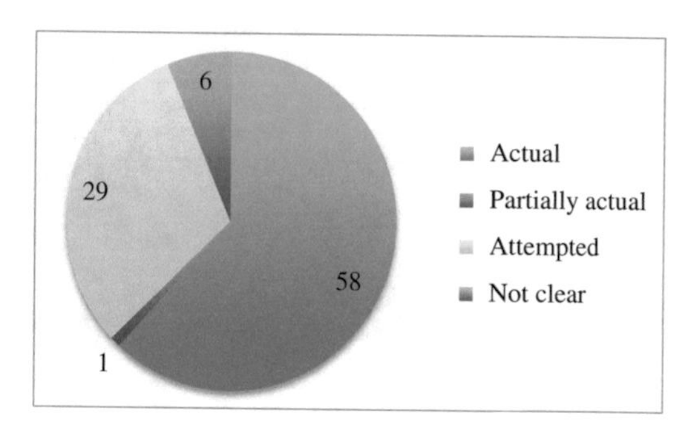

have been recorded on digital media. We asked entrepreneurs to recall their experience to describe the extortion request, to tell us how they reacted and what happened after they reported the event to the police. As regards the kind of extortion request, more than double the extortion cases we have collected from Calabria were actual (58 cases). Only 29 cases were attempted (Fig. 6.5).

As in Sicily, the phenomenon of extortion racket is based on the intimidating power of the mafia-type organisations, which is rooted within a certain territory. In most cases, there is no complaint by the victim of the extortion request. In fact, only 12 people reported the crime to the police. Frequently, the most important information about the crime derives from the lawful interception of communication among members of criminal organisations. As is known, very often the information comes from investigations carried out through video monitoring and environmental interception or by statements of collaborators. Indeed, lawful interceptions of communications have made it possible to gather very important information about the dynamics that characterise the methods adopted by criminal organisations. This information has allowed law-enforcement agencies and those who oppose the criminal phenomenon to achieve important results in the fight against mafia-type organisations.

As regards the type of economic sector affected by the extortion request, 41 were companies belonging to the construction sector, 34 belonged to the trade sector and 19 to other areas, mainly the sector involving the provision of services to businesses and people. Data suggest that in Calabria the figure of a mediator is not used frequently. In most cases the criminal organisation speaks directly to the victims and the possibility of negotiation of the demand is very limited. It should be stressed that among the cases that we have collected a mediator is present only in six cases of extortion, while the element of negotiation is only present in eight cases (Fig. 6.6).

The interviews conducted have also revealed that sometimes those who decide to report the crime are immediately excluded from the market. In these cases, the damage is considerable, as are also the imposition of physical punishments, damages, aggressions, etc. It appears that the most widespread form of extortion in Calabria is defined as one-off and that in relation to the total number of cases of extortion collected, in 83 cases the request was not periodic, while in 8 cases the victims were required to pay una tantum, and not periodically (Fig. 6.7).

As shown by the empirical evidence, occasionally the mafia-type organisations interfere with the management of the company. There are often favours, which

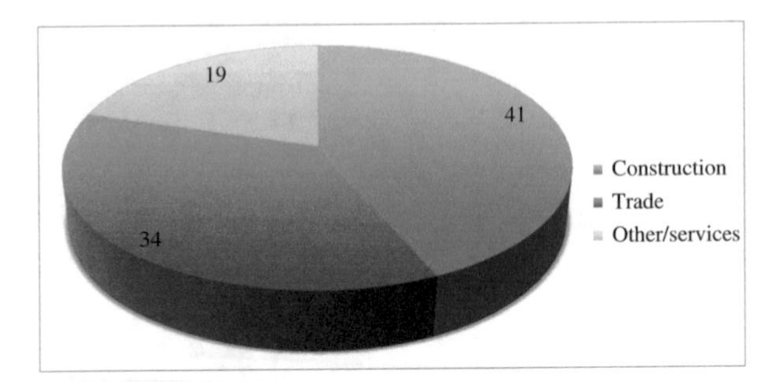

Fig. 6.6 Extortion cases by economic sector

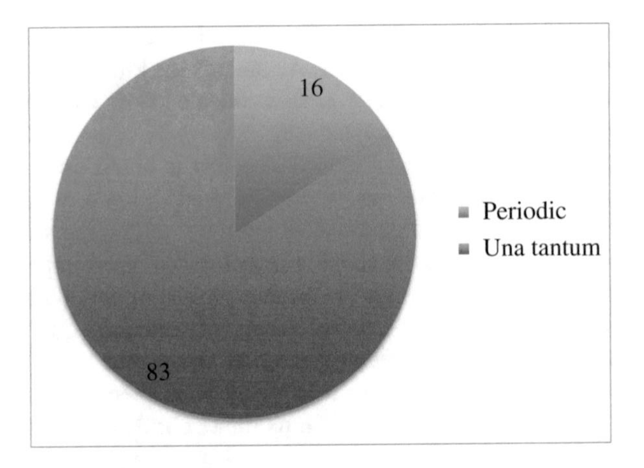

Fig. 6.7 Episodic and periodic demand

sometimes entail to mediate with the employees to solve conflicts. The reference is, in this context, to the different attitudes adopted by the entrepreneur according to which he or she can be acquiescent, resistant or conniving (La Spina, 1999). As is known, the victim of a extortion racket system can (a) decide to pay the required sum. In this case there may be a negotiation; (b) decide not to pay the required sum, (c) pay and never denounce the criminal to the police, pay first and only later report the crime to the police. We also ought to consider the importance of the factors which can induce the subject to report the crime (trust, behaviour of other entrepreneurs, high-profile arrests, economic factors such as shortage of cash or a decline of business, existence of forensic evidence, etc.). As shown in Fig. 6.8, most of the entrepreneurs are acquiescent and most cases of extortion are not discovered as a result of a complaint. The acquiescent entrepreneurs represent the largest group (in relation to our database) and in our opinion this is both a sign of the very strong intimidating power of the organisation and one of the effects of the lack of trust in the institutions. This circumstance is also enhanced by the infiltration of mafia organisations in the public service, as we have seen from the number of municipalities in which the investigations were focused.

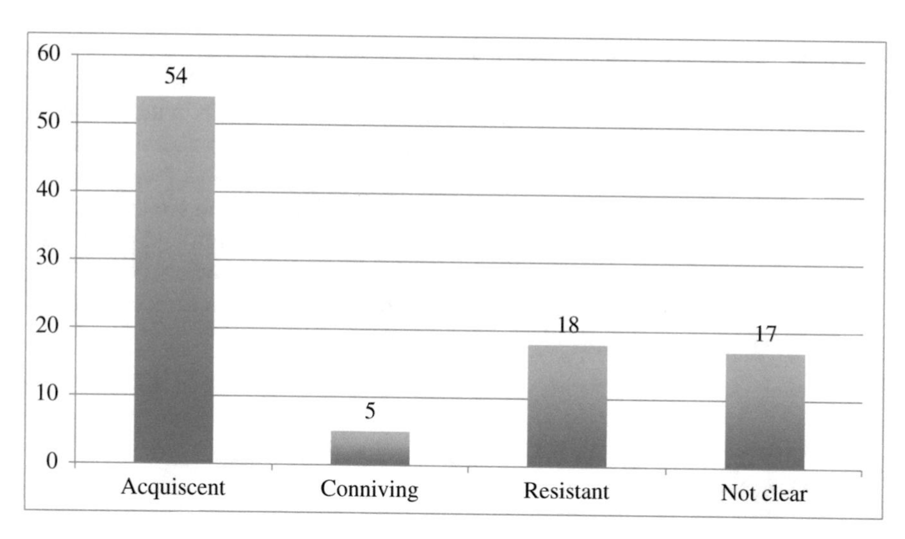

Fig. 6.8 Distribution of victim types

With reference to the decision to report, we can identify in many cases push and pull factors. Interesting information emerged from the stories of witnesses not only in relation to the dynamics of extortion, but also with reference to the forms of interaction and relationships that can be established between mafia and businesses. For example, in some cases, these particular relationships take on the configuration of services that are in various ways linked to the payment of the amount imposed by an entrepreneur. Let us consider a company (A) who is going to invest in the "territory of competence" of the criminal group, where another company (B) is already operating. The representatives of the criminal organisation get in touch, often in a friendly way, with the company that is "in good standing with the payments" and ensure to be able to raise the barriers to entry in the market, perhaps also by requiring the entrepreneur (A) to refrain from making the investment. In these cases, the entrepreneur (who somehow has already spoken to use to their advantage the relationship with the criminal groups) has to deal with a scene from the very ambiguous boundaries, in this case, defining how much the behaviour of the entrepreneur is directly pushed by criminal organisations. The conniving entrepreneur often reports the crime to the police only when their affairs have been revealed. Thus this data has to be analysed with extreme caution. It is very clear, therefore, that conniving business owners rarely complain or claim to have links with criminal organisations. This mainly happens only when they are forced to admit it by strong forensic evidence and cannot act differently. Acquiescent entrepreneurs, on the other hand, only report the extortion if they are discovered or when a police operation reveals the illegal activities. Generally, evidences come out as a result of the complaints of resistant businessmen or through wiretaps that permit to initiate investigations. For this reason, anti-racket associations and the change in attitudes towards organised crime constitute key factors in the contrast of extortion. Similarly, the social representation of mafia-type organisations may influence the behaviour of the victims of extortion. For instance, the way in which the relationship between the suspected or confirmed members of criminal organisations and members of public institutions is considered is a very important factor, because it can contribute in building trust in the state and in the power of police. We may add that the current economic situation, which has greatly reduced the liquidity of companies, is also another factor that correlated with the foregoing, and can encourage the choice to report the crime to the police. The economic conditions of the productive industrial unit and the limited availability of liquidity, as well as the changing attitude of civil society, in combination with the successes achieved by the fight against organised crime, have probably contributed to increase cooperation with the authorities.

6.6 Strengthening Social Responses by Refining Legal Provisions

Some typical social responses against mafia bosses are those of the anti-racket associations, such as Libera or Addiopizzo, which have managed to bring together entrepreneurs, consumers and activists and to strengthen their motivation and ability to resist, providing assistance, advice and tutorship to the victims of mafia crimes as well as to those who receive and manage confiscated proceeds from crime. Other associations which can be remembered are Avviso pubblico, which mobilises civil servants and politicians at the local level, or Ossigeno per l'informazione, which ensures visibility and protection for the journalists threatened by mafiosi. A widespread and growing anti-mafia opposition exists also on a religious level, notably in the Catholic Church (La Spina, 2016b).

The first anti-racket legislation was passed in 1991. The relevant provisions were amended in 1999, because of the difficulties encountered by applicants to benefit from the forms of support provided for by the law. Some other legislative changes were enacted more recently. The number of entrepreneurs who report racketeers appears to be increasing, in the long run (6500 reports of extortion in 2012, not all originating directly from entrepreneurs). However, this number represents only a small fraction if compared to the hundreds of thousands of people who presumably continue to be submissive to extortion. This has prompted to consider the creation of a guarantee fund which could be used to award financial compensation before completion of all required bureaucratic requirements. (Chinnici, La Spina, & Plescia, 2008). Another proposal relates to the introduction benefits aimed at modifying the cost-benefit calculations adopted by acquiescent entrepreneurs (La Spina & Scaglione, 2015, 206ff). It would be desirable, therefore, to amend the reward systems provided by the legislation in order to increase and strengthen the social response against the phenomenon of extortion.

It must also be remembered that some entrepreneurs who pay for the "protection" offered by mafiosi could seem to be subject to an extortion request, but are actually substantially different from the acquiescent ones (be they victims or tame), because they entertain significant collaborative relationships with mafia-type organisations, with a view to earning unnatural benefits, such as the reduction of the market share of their competitors or their complete exclusion from a given market, the imposition of themselves as sole suppliers of certain goods or services, the manipulation and distortion of the market and the management of staff. All these distortions of normal economic activities take place and become possible, due to complicit economic agents "assisted" by the menacing presence of bosses who normally remain in the backstage.

In some cases complicit entrepreneurs were convicted on the basis of their "external support" (concorso esterno) to mafia-type association (art. 416 bis of the Italian Criminal Code). However, this conduct is not easy to prove because evidence must be provided of the significant contribution offered by an economic agent to the maintenance, or at least to the consolidation and strengthening of the association.

Some complicit entrepreneurs were also part of anti-racket groups or have applied for the benefits provided for by anti-racket laws, pretending to be against the bosses while covertly colluding with them. This generates confusion and, at the same time, kind of taints and undermines the social response against organised crime.

In this respect the anti-racket legislation should be modified so as to force non-genuine applicants not only to give back any sum they might have possibly received, but also to punish them with a robust fine (La Spina & Scaglione, 2015, p. 201).

Finally, another proposal relates to the introduction of a new offence, in order to properly tackle the behaviour of those entrepreneurs who have been offered assistance by the mafia in order to avoid competition, and, more generally, in economic transactions. It is not so much the firm which helps the mafia, but rather the mafia-type organisation which supports "trustworthy" economic actors who are at its disposal (La Spina, 2016c). Therefore, on the one hand the existing legislative measures should be further amended in order to adequately combat socially harmful behaviours, while on the other hand strong anti-mafia social responses should also be developed. Honest entrepreneurs who see that their complicit business competitors are punished and possibly excluded from economic activities would then feel reassured and more motivated to work properly, and at the same time would be more confident when considering the option to resist the racket.

References

Chinnici, G., La Spina, A., & Plescia, M. (2008). La resistenza al racket dal punto di vista dell'imprenditoria: Alcune proposte operative. In A. La Spina (a cura di), I costi dell'illegalità. Mafia ed estorsioni in Sicilia. Bologna: Mulino.

Dasgupta, P. (2000), rust as a Commodity, in Gambetta, Diego (ed.) Trust: Making and Breaking Cooperative Relations, electronic edition, Department of Sociology, University of Oxford, chapter 4, pp. 49–72.

Di Gennaro, G., La Spina, A. (2010). I costi dell'illegalità. Camorra ed estorsioni in Campania. Il Mulino, Bologna

La Spina, A. (2008a). Recent anti-mafia strategies: The Italian experience. In D. Siegel & H. Nelen (Eds.), *Organized crime. Culture, markets and policies*. Berlin: Springer.
Gambetta, D. (1993). The Sicilian Mafia: The business of private protection. Cambridge, Mass: Harvard University Press
La Spina, A. (Ed.). (2008b). *I costi dell'illegalità: Mafia ed estorsioni in Sicilia*. Bologna: Il Mulino.
La Spina, A. (guest editor). (2016a). The costs of illegality: Mafia-type organizations and extortion, the case of Campania, special issue of Global Crime, vol. 17, p. 1.
La Spina, A. (2016b). Il mondo di mezzo. Mafie e antimafie. Bologna: Mulino.
La Spina, A. (2016c). Estorsori, estorti, collusi, controllo mafioso dell'economia: Una nuova tassonomia e una proposta di politica del diritto. In Id. E V. Militello (a cura di). Torino: Giappichelli.
La Spina, A., & Scaglione, A. (2015). Solidarietà e non solo. L'efficacia della normativa antiracket e antisura. Soveria Mannelli: Rubbettino.
Matsueda, R. L. (2013). The micro-macro problems in criminology revisited. *The Criminologist, 38*(2), 1–7.
Militello, V., & Siracusa, L. (2010). L'obbligo di denuncia a carico dell'imprenditore estorto fra vecchi e nuovi paradigmi sanzionatori, in I costi dell'illegalità. Camorra ed estorsioni in Campania, di Gennaro-La Spina (cur.). Bologna: Il Mulino, 483s.
Monzini, P. (1993). L'estorsione nei mercati leciti e illeciti. Libero Istituto Universitario Carlo Cattaneo.
Punzo, V. (2013). Simulazione sociale e spiegazione generativa del crimine. Applicazioni allo studio del fenomeno mafioso. In A. La Spina et al. (Eds.), *Mafia sotto pressione* (pp. 99–126). Milano: FrancoAngeli.
Santino, U. (2006). *Dalla Mafia alle mafie. Scienze Sociali e Crimine organizzato*. Soveria Mannelli: Rubbettino.
Scaglione, A. (2008). Il racket delle estorsioni. In A. La Spina (Ed.), *I costi dell'illegalità: Mafia ed estorsioni in Sicilia* (pp. 77–112). Bologna: Il Mulino.
Sciarrone, R. (2009). *Mafie vecchie, mafie nuove*. Roma: Donzelli.
Sos Impresa-Confesercenti. (2007). Le mani della criminalità sulle imprese, X Rapporto, Roma, disponibile all'. http://www.sosimpresa.it/iniziative/2007/assemblea2210/decimo_rapporto.pdf
Transcrime. (2013). Joint research center on transnational crime. Progetto PON SICUREZZA 2007–2013 Gli Investimenti delle Mafie, Universita Cattolica del Sacro Cuore di Milano.
Vaccaro, A. (2012). To pay or not to pay: Dynamic transparency and the fight against the mafia's extortionists. *Journal of Business Ethics, 106*, 23–35.

Chapter 7
An Agent-Based Model of Extortion Racketeering

Luis G. Nardin, Giulia Andrighetto, Áron Székely, Valentina Punzo, and Rosaria Conte[†]

7.1 Introduction

Mafias can be considered as criminal organisations that are in the business of producing, promoting, and selling protection. Put simply, they are protection racketeering groups (Gambetta, 1993). They are widespread across the globe, among them are the Russian mafia (Varese, 1996, 2001), the Yakuza (Hill, 2006), the Triads (Morgan, 1960), and the Sicilian mafia (Savona, 2012).

They cause both economic and social damage to the societies in which they are embedded (Daniele, 2009). One reason is because they do offer their services not only to people and businesses that participate in legal transactions, but also—and likely more so—to those who are involved in illegal transactions, allowing markets for these illegal, and frequently harmful, goods and services, to exist (Gambetta,

---

Parts of this chapter were published in Nardin et al. (2016). Nardin, L. G.; Andrighetto, G.; Conte, R.; Székely, Á.; Anzola, D.; Elsenbroich, C.; Lotzmann, U.; Neumann, M.; Punzo, V. & Troitzsch, K. G. (2016). 'Simulating protection rackets: A case study of the Sicilian mafia'. *Autonomous Agents and Multi-Agent Systems*, Online, 1–31. Reprinted with permission of the Journal of Autonomous Agents and Multi-Agent Systems.

[†]Author was deceased at time of publication.

L.G. Nardin
Institute of Cognitive Sciences and Technologies (ISTC), Italian National Research Council (CNR), Via Palestro, 32, Rome 00185, Italy

Schuman Centre for Advanced Studies, European University Institute, Fiesole, Italy
e-mail: gnardin@gmail.com

G. Andrighetto (✉)
Institute of Cognitive Sciences and Technologies, Italian National Research Council (CNR), Rome, Italy

Schuman Centre for Advanced Studies, European University Institute, Fiesole, Italy
e-mail: giulia.andrighetto@istc.cnr.it

C. Elsenbroich et al. (eds.), *Social Dimensions of Organised Crime*, Computational Social Sciences, DOI 10.1007/978-3-319-45169-5_7

1993, pp. 226–244). They can also enforce cartels among businesses, driving up costs, hurting consumers, and reducing productivity (Gambetta, 1993, pp. 195–225; Varese, 2013 p. 5). Moreover, they often seek to establish and distort the political and institutional processes. One study estimates that the mafias in Italy combined produce tax-free capital that was equivalent to about 7 % of the national GDP in 2007 (Barone & Narciso, 2013). Other studies have examined the economic harm caused by the Italian mafias, and organised crime more generally, and find that their presence substantially hampers economic growth (Lavezzi, 2008; Pinnotti, 2015a, 2015b).

Thus, overcoming or at least limiting protection rackets is a highly desirable policy objective. Yet, this is a difficult task since buyers actively seek out the protection provided by some groups and, if not, the threat of economic or physical violence and norms of secrecy and honour can dissuade others from cooperating with the police. Hence, protection racketeers receive the support from portions of society and implicit protection from others by their refusal to cooperate.[1]

An important step to take in countering protection racketeering groups is to deepen our understanding of them. These groups, however, are notoriously difficult to investigate. Apart from the obvious risks that adventurous empirical researchers face, there is a more fundamental issue. Even those willing to overlook (or able to elude) the potential danger cannot avoid the secretive nature of such groups that hide their criminal activities from prying eyes making it difficult to uncover empirical data about their operations and dynamics. Even the empirical data that are extracted—the judicial documents from the Maxi Trial (Alfonso, 2011) are one example—capture only a certain proportion of the true levels of the criminal activities, and, in any case, they are not beyond reproach because they may be biased in ways that are difficult to correct for: captured members may not be representative of the group (they are the *losers*) or they may have incentives to distort their testimony. Additionally, unlike many other types of crime, the victims often have little incentive to come forward, in part, because of the long-term, semi-collusive nature of protection rackets.

Such hindrances can be, in part, alleviated with simulation models. They can function as key tools that provide a data source with which to compare or enrich empirical data, bolstering or conflicting with what has already been found. In this sense, such models can be used as checks for what has been found providing further reassurance in case there is congruence, or as warning flags that highlight questionable data when incongruence occurs.

[1] Another part of this is likely down to a selection effect in that those criminal groups which are not entrenched in their milieu do not survive.

Á. Székely
Institute of Cognitive Sciences and Technologies, Italian National Research Council (CNR), Via San Martino della Battaglia 44, Rome 00185, Italy
e-mail: aron.szekely@istc.cnr.it

V. Punzo
University of Palermo, Palermo, Italy
e-mail: valentinapunzo@libero.it

Ultimately these efforts to model protection racketeering should not only help us to understand how such groups work, but also enrich our knowledge of how to stop them working. Simulation models can and should also be used to test bed anti-racket policies. Two important anti-racket approaches can be called *legal* and *social norm*-based approaches (see Chap. 4 – 6). In the legal norm-based approach, the state uses legal norms, or laws, that are norms issued by legal authorities and enforced by specialised actors (Elster, 2007, p. 357). In the social norm-based approach, various actors, be it the state, non-governmental organisations (NGOs), or citizens' groups, try to change peoples' actions through non-legal means, targeting social norms in particular by shaping their expectations and beliefs about what is socially appropriate. We can define social norms as socially shared rules that prescribe what individuals ought or ought not to do and that are often spontaneously monitored and enforced by peers (Bicchieri, 2006, Conte, Andrighetto, & Campennì, 2013, Elster, 2009). Campaigns, discussions, and information spreading, all lacking the bite of the law, are nevertheless powerful tools for behaviour change. Social norms are both a social and a cognitive phenomenon undergoing complex dynamics (Conte et al., 2013, Conte & Castelfranchi, 1999). They influence people by shaping their mental representations, such as normative beliefs and normative goals, which can subsequently affect their behaviour.

Agent-based modelling (ABM) is a computational modelling approach that is particularly suited for studying dynamics that integrate cognitive and social aspects as it allows agents to be influenced by macro-level social factors, explicitly represent these as mental constructs at the micro-level, and consequently reconstitute the social reality via their actions. Here, we describe the *Palermo Scenario*[2] (Nardin et al., 2016), an agent-based model of protection rackets aimed to deepen our understanding of protection rackets, and help policymakers to evaluate methods for destabilising them. Additionally, since the system is explicitly specified, we can use it to investigate the entire causal pathway from cause to effect: not only from actions to mafia destabilisation, but also the intermediate actions along the path and actors' internal mental representations among the population.

This chapter unfolds as follows. In Sect. 7.2, the Palermo Scenario, along with its main actors and their decision-making, social norms, and dynamics, is described. The description of how the social norms influence on the actors' decisions is given in Sect. 7.3.

7.2 Palermo Scenario

Based on empirical evidence extracted from a range of sources (see Sect. 12.2 and Chap. 6), as well as discussions with GLODERS stakeholders, who are actively involved in anti-mafia policies or initiatives, and members of the GLODERS project,

[2] The model is denominated by Palermo Scenario because most of the empirical data used to develop the model was collected in the area of Palermo. Despite its name, it is worth noting that the model is flexible enough to represent the dynamics behind other racketeering groups.

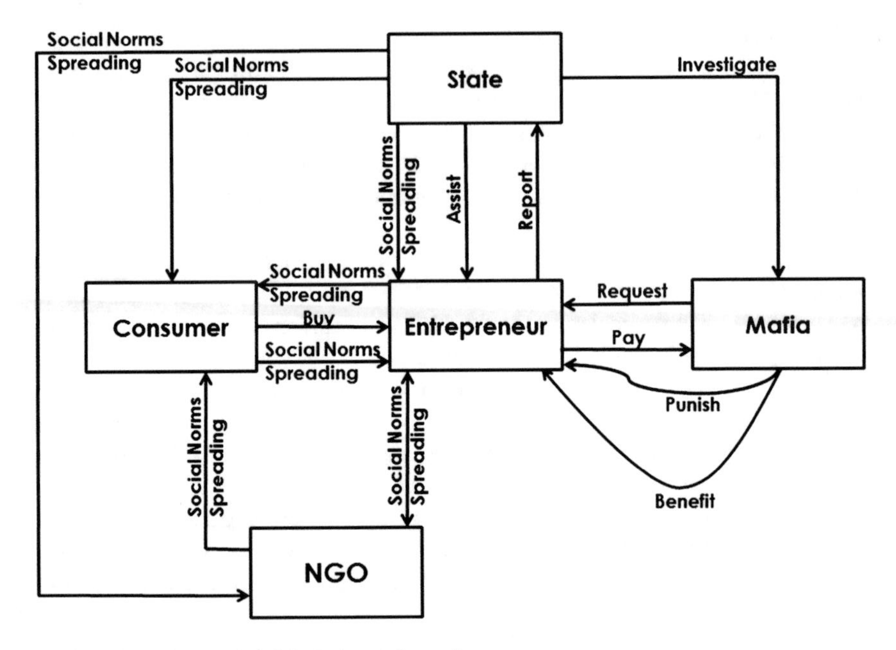

Fig. 7.1 Interrelationship of the Palermo Scenario actors

we identified five key actors in the dynamics of the mafia phenomenon and their interrelationships (shown in Fig. 7.1): Entrepreneurs, Consumers, the State, the mafia, and a Non-Governmental Organisation.[3] Notice that this is a stylised fact simulation model; thus we adopt several simplifications that nevertheless capture the main characteristics of the mafia phenomenon. We start by going through each actor.

Entrepreneurs represent businessmen and liberal professionals. They are modelled as multiple agents and are the central actors in the model. They sell products to consumers at a range of prices and receive income, and make a number of decisions using a combination of economic and normative reasoning. Entrepreneurs can

- Decide to pay pizzo if approached by Mafiosi
- Report pizzo requests to the State if they decide not to pay pizzo
- Report to the State damages that they sustained from mafia attacks
- Collaborate with the State against specific Mafioso if approached by the State
- Join the Non-Governmental Organisation, thereby signalling that they are unwilling to pay pizzo, likely to report pizzo requests and mafia punishments, and obtain respite from mafia requests

The State represents the government institutions. It can

- Imprison Mafiosi: Mafiosi can be sent to prison after investigation by the police, who work with either specific evidence obtained from entrepreneurs or evidence

[3] These sources are judicial documents, confiscated mafia documents such as *Libri Mastri* (accounting books used by some Mafiosi to record various information about extortion payers and that are occasionally discovered by the police), academic studies, literature, and other sources such as newspapers and television interviews.

obtained from general day-to-day observation and police activity. Naturally, investigations based on specific evidence are more effective than those based on general observation. After the police captures a Mafioso, the police may find information about the Entrepreneurs who paid pizzo to that Mafioso: the Mafioso may provide information (i.e. *pentiti*[4]) or the information may be found in assorted documents such as *Libro Mastro*. The State can then use this evidence to elicit collaboration from those Entrepreneurs by threatening them with punishment and if collaboration is obtained the State uses their information to increase the possibility of prosecuting that Mafioso.
- Support Entrepreneurs who have suffered damages at the hands of Mafiosi: Entrepreneurs who have suffered some damages from mafia retaliation can apply for monetary support to a fund that is set up specifically for this purpose, the *Fondo di Solidarietà* (i.e. a state-run fund to support mafia victims), which contains resources that depend on a politically determined component and a component derived from the resources of captured Mafiosi.
- Spread facts about successful actions that it has carried out against the mafia (consider this as the State providing information to journalists who report and propagate the news in newspapers and television programmes).
- Change peoples' attitudes regarding the mafia using campaigns and education regarding appropriate behaviour, some of which is done by sponsoring and supporting anti-racket festivals, such as the *Festival della Legalità*, or by promoting the culture of legality.

The mafia represents criminal organisations. It is composed of many actors who

- Request pizzo from Entrepreneurs
- Provide benefits to paying Entrepreneurs (e.g. protection from predation, and contract and cartel enforcement)
- Punish non-paying and reporting ones with a specific severity. They are coordinated in their actions—whom they target, how often they request pizzo, how much they request, and how severely they punish—because they are part of the same family. Mafiosi can
- Turn pentiti (a very unlikely event) and help the State capture other Mafiosi
- Mafiosi who are captured by the State are temporarily removed from the simulation and may provide information about other Mafiosi and the Entrepreneurs who paid pizzo to it in the past allowing the State to approach these Entrepreneurs for evidence.

Consumers are multiple actors who do not directly interact with the mafia. They are connected to other Consumers and Entrepreneurs in a social network; this determines the other actors with which they socially interact. Consumers have the goal to purchase a product and their single decision is to buy a product from Entrepreneurs. The decision regarding which Entrepreneur to buy from is based on a combination of economic considerations (i.e. price of the product) and normative considerations (i.e. relative strength of the norm of buying from Entrepreneurs who do not pay

[4] *Pentiti* designate former members of criminal organisations that, in most cases following their arrest, decide to collaborate with the judicial system to help investigations.

pizzo, dynamically updated over the simulation). They serve as reservoirs of normative attitudes and behaviours and automatically spread information that can influence other Consumers and Entrepreneurs.

The Non-Governamental Organisation is a single actor that embodies a civil society or business organisation. It promotes the culture of legality among Entrepreneurs and Consumers through events such as talks in schools, or the organisation or participation in festivals: for instance, the civic organisation *Libera* is the main organiser for the aforementioned Festival della Legalità. It serves as an organisation that Entrepreneurs can join if they are not paying pizzo.

7.2.1 Decision Processes

The decision-making of actors in the Palermo Scenario can be broadly divided into two different levels of complexity. Entrepreneurs and Consumers are endowed with more sophisticated decision-making abilities and base their choices on a combination of economic and social norm-based reasoning, whereas the State, the mafia, and the Non-Governmental Organisation are represented as reactive actors whose decisions are defined exogenously based on fixed probabilities specified at the start of the simulation.

The Entrepreneurs' and Consumers' decisions are taken assuming that the utility of an actor consists of an *individual* component, which represents the economic part of their reasoning, and a *normative* component, which represents the social norm-based aspect. The individual component approximates instrumental decision-making and involves strict cost-benefit calculations that motivate actors to take decisions that maximise their own direct utility, independently of what a certain norm dictates. The normative component models the actor's motivation to comply with a norm. It is a function of *norm salience*, a parameter updated by each actor based on its own behaviour and the information gathered by observing the behaviour of other actors.

Following Conte et al. (2013, p. 99), we use *norm salience* to refer to a measure that indicates how active and prominent, or inactive and inconspicuous, a norm is within a group in a given context. Formally,

$$\mathrm{Sal}^{n} = \frac{1}{\alpha}\Big(\beta + \Big(\frac{C-V}{C+V}\times w_c + \frac{O_c - O_v}{O_c + O_v}\times w_o + \frac{\max(0,(O_v+V)-P-S)}{O_v+V}\times w_{npv}$$
$$+\frac{P\times w_p + S\times w_s}{\max(P+S, O_v+V)} + \frac{E_c - E_v}{E_c+E_v}\times w_e\Big)\Big)$$

where n is the norm being evaluated; α and β are normalisers that render the norm salience value in the range [0,1]; C is the number of times the actor complied with the norm n; V is the number of times the actor violated the norm n; O_c is the number of times the actor observed other actors complying with the norm n; O_v is the

Table 7.1 Social cues and weights for the norm salience updating (Andrighetto et al., 2010)

Cue	Description	Weight
C/V	Own norm compliance/violation	$w_c = (+/-)\ 0.99$
O	Observed norm compliance	$w_o = +0.33$
NPV	Non-punished violators	$w_{npv} = -0.66$
P	Observed/applied/received punishment	$w_p = +0.33$
S	Observed/applied/received sanction	$w_s = +0.99$
E	Observed/applied/received norm invocation	$w_e = +0.99$

Table 7.2 Summary of the social norms in the Palermo Scenario

Actor	Id	Social norm
Entrepreneur	N^P	Pay pizzo request
	N^{NP}	Do not pay pizzo request
	N^R	Report pizzo request
	N^{NR}	Do not report pizzo request
Consumer	N^{NB}	Avoid paying pizzo Entrepreneurs

number of times the actor observed other actors violating the norm n; P is the number of punishments received, applied, or observed due to the violation of norm n; S is the number of sanctions received, applied, or observed due to the violation of norm n; E_c is the number of messages received from others 'demanding' that the actor complies with the norm n; and E_v is the number of messages that the actor received 'demanding' the violation of the norm n.

Each term in the norm salience calculation has a weight value associated with it, and the coefficients α and β have the values 6.27 and 2.97, respectively. The weights are used to assign different importance to each of the factors in generating the overall norm salience. In Table 7.1, the weight associated to each term is presented, the values of which are based on the work of Cialdini, Reno, and Kallgren (1990). It is important to stress that the important aspect of these weights is the proportionality among them and not their specific value.

Entrepreneurs use these economic and normative aspects to decide whether or not to pay or report pizzo request to the State. Consumers use it to decide which Entrepreneur to purchase a product from. Those decisions are intimately related to the social norms modelled in the Palermo Scenario described next.

7.2.2 *Social Norms*

The summary of the specific social norms that Entrepreneurs and Consumers consider in the Palermo Scenario is shown in Table 7.2. For a discussion of social norms in protection rackets, please refer to Chap. 4.

N^P and N^{NP} are norms that potentially influence the decision of Entrepreneurs to pay pizzo to Mafiosi following a request, and N^R and N^{NR} are norms that can play a role in Entrepreneurs' decision to report the request for pizzo by Mafiosi to the State. N^{NB} is a norm that can influence the Consumers' decisions regarding which Entrepreneur to purchase a product from.

Norms N^{P} and N^{NR} are part of the set of norms that are associated with the traditional mentality of the individuals regarding the mafia, in which pizzo should be paid and not reported to the police (*omertà*). Conversely, norms N^{NP} and N^{R} represent the set of norms that correspond to a recent emerging anti-racket sentiment that is based on the understanding of the social and economic harm caused by the mafia. Differently to these, norm N^{NB} is one factor that is used by Consumers to rank the different Entrepreneurs that may buy a product from.

7.3 Interplay of Social Norms and Decision Processes

We now go over the racket-related social norms that we identified in Sect. 7.2.2, and verbally describe them from the perspective of the relevant decision of each actor.

Entrepreneurs recognise and consider four different social norms. These norms are to (1) pay pizzo, (2) do not pay pizzo, (3) do not report pizzo requests to the police, and (4) report pizzo requests to the police.

The first two relate to the decision of the Entrepreneur to pay pizzo, following a pizzo request by a Mafioso. One prescribes that the Entrepreneur should pay pizzo while the other proscribes the action and entails that the Entrepreneur should not pay pizzo. The second two norms relate to the Entrepreneur's decision to report a pizzo request to the police or not: a decision that is taken by Entrepreneurs if they chose not to pay following a pizzo request.

Entrepreneurs simultaneously hold both the norm proscribing action X and the norm prescribing that same action X. More than just an absence of a rule that prescribes that action, in the converse norm, the action is actually proscribed, and may be enforced through sanctions. For the social norm to pay pizzo, this means that there are reciprocal expectations about paying pizzo. For the social norm to not report pizzo requests, the same pertains, although here there is some evidence that it can be enforced via sanctions. While it may seem odd to have a social norm prescribing pizzo payment and proscribing reporting, there are real-life examples in which people are punished in some way, ostracised for instance, for violating them. In one case, citizens boycotted a shopkeeper because he reported pizzo requests to the police (Diliberto, 2013).

Entrepreneurs may be approached by Mafiosi and asked to pay pizzo, and subsequently, they have to decide whether to pay or not. From the perspective of the Entrepreneur, two social norms are relevant to making this decision: the norm to 'pay pizzo' and the norm 'do not pay pizzo'. Let 'Sal^{P}' indicate the norm salience of former and 'Sal^{NP}' the norm salience of the latter, and 'NG' the normative goal that is adopted for this decision.

The norm salience values for the two norms are generated and updated during the simulation. When the Entrepreneur is faced with the decision, it compares the values of the two norm saliences ('Sal^{P}' and 'Sal^{NP}'). If the norm salience of pay is higher than the norm salience of not pay, then the Entrepreneur adopts the norm of paying as its normative goal. So, if $Sal^{P} > Sal^{NP}$, then $NG = Sal^{P}$. When paying is adopted as the goal, the eventual probability of paying is increased. Otherwise, if the norm salience of not paying is higher than the norm salience of paying, then the Entrepreneur adopts not paying as its goal. More specifically, in this case, the Entrepreneur uses one minus the norm salience value of not paying as its normative goal. So, if $Sal^{NP} \geq Sal^{P}$, then $\text{NG} = 1 - Sal^{NP}$. This is implemented in such a way because it ensures that a higher not-pay norm salience leads to a lower probability of paying.

The Entrepreneur then weights and combines the normative goal value that it adopts with a weighted value for its other goal, the 'individual goal' relevant for this decision, and out of these creates a threshold value 'T^*'. While we do not discuss the individual goal here, it is relevant to mention that the individual goal varies according to which decision the Entrepreneurs are making. A randomly selected number is drawn from the interval 0–1, and if this number is less than the threshold then the Entrepreneur pays; otherwise it does not (i.e. if the number selected from $[0, 1] < T^*$ then pay; otherwise do not).[5]

Consider now the decision to report pizzo requests to the police or not to do so. Only Entrepreneurs who previously chose not to pay pizzo face this decision. The two social norms relevant to this decision are 'report pizzo request' and 'do not report pizzo request'; let their respective norm saliences be indicated by 'Sal^{R}' and 'Sal^{NR}' and 'NG' the normative goal that is adopted for this decision.

When deciding, Entrepreneurs compare the norm saliences attached to the two norms, 'Sal^{R}' and 'Sal^{NR}'. If the salience of the norm to report is greater than the salience of the norm not to report, then that norm, and associated value, is adopted as the Entrepreneurs's goal. Thus, if $Sal^{R} > Sal^{NR}$ then $NG = Sal^{R}$. Otherwise, if the salience of the norm not to report is greater than the salience of the norm to report, then the Entrepreneur adopts the normative goal of not reporting. Specifically, one minus the salience value of the not-report norm. If $Sal^{NR} \geq Sal^{R}$, then $NG = 1 - Sal^{NR}$.

The normative goal that is adopted is then combined, in a weighted manner, with the individual goal relevant for this decision, and a threshold 'T^{**}' is created. A number is then randomly selected from the interval 0–1; if the number is less than the threshold then the Entrepreneur decides to report; otherwise it does not (i.e. if a randomly chosen number from $[0, 1] < T^{**}$ then report; otherwise do not).

The final decision of Entrepreneurs that is affected by their social norms is their decision to join the *Organisation* or not. This is a decision that can be taken by

[5] Consider a high norm salience for the norm 'do not pay pizzo' and assume that it is adopted. In this case, since $1 - Sal^{NP}$ is used, the threshold that emerges from the combined goals is low, meaning that the probability that a randomly drawn number is greater than T^* is high. Therefore, the probability of paying is low. In contrast, consider a low norm salience for 'do not pay pizzo' and assume that it is adopted. In this case, the normative goal value is high (since $1 - Sal^{NP}$), and thus, the probability that the threshold will be exceeded is low, and consequently the probability of paying is high.

Entrepreneurs only after a pizzo request and only if the Entrepreneurs decided not to pay—no 'fakers' can join the Non-Governmental Organisation.

This decision employs the norm salience value that is created, and updated, for the norm to report pizzo requests: 'Sal^{R}'. This norm's salience is more stringent than that for not paying pizzo, in the sense that it is harder to achieve a higher value, because the Entrepreneur will hardly ever observe others following it and thus is the one that most closely corresponds to the idea of extreme indignation arising from anti-Mafiosi sentiment. The threshold used in this decision is exogenously set; let this threshold be represented by 'NG^*'. If the salience for the norm to report is higher than this threshold, then the entrepreneur joins. Alternatively, if the salience for this norm is lower than the threshold, the Entrepreneur does not join (i.e. if $Sal^{R} > NG^*$ then join while if $Sal^{R} \leq NG^*$ then do not join). For this decision, there is no individual goal—Entrepreneurs do not combine the normative goal with an individual goal—because empirically Entrepreneurs seem to be motivated by normative reasons and not instrumental ones.[6] Joining is irreversible, so Entrepreneurs who join cannot leave.

Consumers recognise a single social norm: (1) do not buy from pizzo-paying shops. And have one decision that is affected by their social norms: their *purchasing decision*. The social norm that is relevant to this decision is 'do not buy from pizzo-paying shops'. Let the salience for this norm be represented by 'Sal^{NB}'.

A set of Entrepreneurs is randomly selected for consideration by the Consumers. They are the Entrepreneurs from whom the Consumer may wish to buy a product. The norm salience for avoiding pizzo-paying shops is then integrated into a ranking formula that also considers the price of the product sold by each Entrepreneur. Each Consumer is consequently left with a list of ranked Entrepreneurs, and the consumer chooses the highest ranked Entrepreneur. Consumers with higher norm saliences, 'Sal^{NB}', rank shops that they believe to be paying pizzo further down the list. They form their beliefs about each shop's pizzo payment based on the reputation of each shop, which in turn is based on observations of shop behaviour and sanctions applied to those shops by others.

Consumers and Entrepreneurs update the salience of their norms throughout the simulation. They both consider their own history of compliances or violations, the history of others' compliances or violations, the history of punishments and lack of punishments that occurred following their own and others' norm violations, the history of sanctions that occurred following their own and others' norm violations, and explicit norm invocations to comply or violate the norm that they receive.

The State holds two legal norms: (1) combat the mafia and (2) assist Entrepreneurs. These are based on the *Rognoni-La Torre Law n. 646 of 13/9/1982* that introduced into the Italian criminal code the crime of mafia-style criminal organisation (art. 416 bis) and the possibility of confiscating mafia properties with their consequent social reuse. In addition, *Law n. 8 of 15/01/1991* and *Law n. 82 of 15/03/1991* aim at providing denouncing incentives and protecting victims who report extortion activities.

[6] Although in theory they can be motivated to join for instrumental reasons, we did not implement this due to the unnecessary complexity that would be added to the model.

Finally, *Law n. 44 of 23/02/1999* and *Law n. 512 of 22/12/1999*, respectively, introduced economic support to victims of extortions and the solidarity fund for victims of Mafioso-style crimes and intimidation. These norms are implemented as actions that the State carries out. The State does not hold any social norms. However, it can promote the social norms held by Entrepreneurs and Consumers.

The State combats the mafia using two different types of investigations. Police officers conduct general investigations on an ongoing basis, keeping a general lookout for pizzo requests and punishments enacted by Mafiosi. It also carries out specific investigations that are based on reports by Entrepreneurs. In addition to such direct anti-racket legal norms, the State spreads normative information, exhorting Entrepreneurs and Consumers to pursue actions that undermine the mafia, and it spreads information about successful anti-racket operations that it carried out.

Regarding assistance to Entrepreneurs, the State has a resource pool, partly comprised of resources confiscated from the mafia and partly composed of money allocated into it by the government: the Fondo di Solidarietà. Entrepreneurs who report Mafiosi activity and are punished for doing so can obtain reimbursement from the fund.

Generally put, legal norms structure interactions—with the sole exception of Entrepreneurs' decision to collaborate—while social norms influence agents' decision-making within interactions.

References

Andrighetto, G., Villatoro, D., & Conte, R. (2010). Norm internalization in artificial societies. *AI Communications, 23*, 325–339. IOS Press.

Alfonso, G. (2011). *Il maxiprocesso venticinque anni dopo* (Memoriale del presidente). Rome: Bonanno.

Barone, G., & Narciso, G. (2013). *The effect of mafia on public transfers*. The Rimini Centre for Economic Analysis.

Bicchieri, C. (2006). *The grammar of society: The nature and dynamics of social norms*. New York: Cambridge University Press.

Cialdini, R. B., Reno, R. R., & Kallgren, C. A. (1990). A focus theory of normative conduct: Recycling the concept of norms to reduce littering in public places. *Journal of Personality and Social Psychology, 58*(6), 1015–1026.

Conte, R., Andrighetto, G., & Campennì, M. (Eds.). (2013). *Minding norms: Mechanisms and dynamics of social order in agent societies* (Oxford series on cognitive models and architectures). Oxford: Oxford University Press.

Conte, R., & Castelfranchi, C. (1999). From conventions to prescriptions. Towards an integrated view of norms. *Artificial Intelligence and Law, 7*, 119–125.

Daniele, V. (2009). Organized crime and regional development. A review of the Italian Case. *Trends in Organized Crime, 12*(3–4), 211–234.

Diliberto, P. (2013). Addiopizzo 2.0. *Il Testimone*. MTV.

Elster, J. (2007). *Explaining social behavior: More nuts and bolts for the social sciences* (2 revth ed.). Cambridge, MA: Cambridge University Press.

Elster, J. (2009). Social norms. In P. Hedström & P. Bearman (Eds.), *The Oxford handbook of analytical sociology*. Oxford: Oxford University Press.

Gambetta, D. (1993). *The Sicilian mafia: The business of private protection*. Cambridge, MA: Harvard University.

Hill, P. B. E. (2006). *The Japanese mafia: Yakuza, law, and the state*. Oxford: Oxford University Press.

Lavezzi, A. M. (2008). Economic structure and vulnerability to organised crime: Evidence from Sicily. *Global Crime, 3*, 198–220.

Morgan, W. P. (1960). *Triad societies in Hong Kong*. Hong Kong: The Government Printer.

Nardin, L. G., Andrighetto, G., Conte, R., Székely, Á., Anzola, D., Elsenbroich, C., et al. (2016). Simulating protection rackets: A case study of the Sicilian mafia. *Autonomous Agents and Multi-Agent Systems*, Online, 1–31.

Pinnotti, P. (2015a). The economic costs of organised crime: Evidence from Southern Italy. *The Economic Journal, 125*(586), F203–F232.

Pinnotti, P. (2015b). The causes and consequences of organised crime: Preliminary evidence across countries. *The Economic Journal, 125*(586), F158–F74.

Savona, E. U. (2012). Italian mafias' asymmetries. In D. Siegel & H. van de Bunt (Eds.), *Traditional organized crime in the modern world* (Vol. 11, pp. 3–25). Berlin: Springer.

Varese, F. (1996). What is the Russian mafia? *Low Intensity Conflict and Law Enforcement, 5*, 129–138.

Varese, F. (2001). *The Russian mafia: Private protection in a new market economy*. Oxford: Oxford University Press.

Varese, F. (2013). *Mafias on the move: How organized crime conquers new territories*. Princeton, NJ: Princeton University.

Chapter 8
Extortion Rackets: An Event-Oriented Model of Interventions

Klaus G. Troitzsch

8.1 Introduction

This chapter documents the final NetLogo (Tisue & Wilensky, 2004) version of the Palermo case study simulation model.[1] Its predecessors were originally prepared as rapid prototypes to prepare and to instruct the Java-based model but soon developed into a flexible tool with a graphical user interface. The predecessors have been described in a number of publications (Troitzsch, 2015a, 2015b, 2016a, 2016b)—they were designed as period-oriented simulations, i.e. in every period agents made their decisions and performed their actions, usually several in a row, e.g. for an extorter the action of approaching a potential victim and taking its money, being denounced, prosecuted and finally put to custody and convicted. For the purpose of serving as a guide for the GLODERS-S simulator (Nardin et al., 2016) and for sensitivity analyses this was sufficient, but it turned out to be a matter of a few days of programming to turn the period-oriented NetLogo model into an event-oriented one with the help of the time extension (Sheppard & Railsback, 2014).

As GLODERS-S and the period-oriented NetLogo model, the event-oriented version consists of four types of agents: enterprises or shops, extorters, police and consumers. Unlike GLODERS-S, the extorters are independent and can form families; the police officers are independent, too, i.e. both extorters and police officers can make decisions according to the individual norms they learnt during the simulated history. Hence for all types of agents the salience of norms is computed before any action decision is taken.

[1] The model can be downloaded from http://ccl.northwestern.edu/netlogo/models/community/EONOERS.

K.G. Troitzsch (✉)
Computer Science Department, Universität Koblenz-Landau,
Universitätsstraße 1, Koblenz, Rheinland-Pfalz 56070, Germany
e-mail: kgt@uni-koblenz.de

C. Elsenbroich et al. (eds.), *Social Dimensions of Organised Crime*,
Computational Social Sciences, DOI 10.1007/978-3-319-45169-5_8

Table 8.1 Agent type, norms and actions

Agent type	Norms	Actions
Shops	Denounce-extortion	Denounce
	Do-not-denounce	
	Do-not-pay-extortion	Pay
	Pay-extortion-as-everybody-does	
Consumers	Do-not-shop-at-extortion-payer	Switch to Addiopizzo shop
	Buy-from-extortion-payer	
Police	Anxiety-is-justified	Prosecute
	Try-hard-to-imprison	
Extorters	Do-not-betray-colleagues	Become pentito
	Abjure-crime	

8.2 The Repertoire of Norms and Actions

The actions and the behaviour of the four types of agents are controlled by a number of norms whose salience is continuously updated by observations of norm-related behaviour and particularly of norm invocations issued by other agents of the same or of a different type. The agent types, their norms and their actions are listed in Table 8.1. The salience of a norm is calculated according to a formula discussed in Sect. 7.3 which converts the counters of norm-related observations and invocations into a number between 0 and 1. The saliences of the two norms relevant for an action are in turn converted into a normative drive to perform this action and finally mixed with the individual drive calculated from the utility of the action, and this weighted sum of normative and individual drive is taken as the probability to perform the respective action in the current situation.

So far the period-oriented and the event-oriented versions are equal (and use the same code) but the event-oriented version needs much more sophistication for scheduling events in a reasonable order. The actions of agents which trigger events starting new actions of the same agent or an agent of another type are listed in Table 8.2 together with the delay between the triggering and the triggered actions (which is usually a random number of hours or years[2]).

Table 8.2 also gives a nearly complete overview of the actions which can be taken by the agents of the different types (some of these actions use additional procedures to describe what the agent will have to do in order to perform them). Besides

[2]The standard duration of a run is 2 years, which—for 100 shops, 50 extorters, 10 police and 800 customers, some 2000 extortion attempts and 37,000 exchanged invocation messages—takes a standard PC with eight processors about 2 min on an average in batch mode (a single run takes 20–30 min). Three times as many agents of each type with about 5000 extortion attempts and about 100,000 messages exchanged means about 30 min per run. The delays mentioned in Table 8.2 are not calibrated against any empirical data as even the Sicily and Calabria database (see Chap. 6) does not contain sufficient details for such a calibration. The simulated time units can only have a rough correspondence to real-time units.

Table 8.2 Scheduling of events

Acting agent	Controlling action	Condition	Activated agent	Triggered action	Delay
Court	Convict	Sentenced	Extorter	Leave-jail	3–8 years
Court	Convict	Acquitted	Extorter	Become-active	24 h
Extorter	Become-active	Met nobody	Extorter	Become-active	730 h
Extorter	Return-to		Shop	Wait-for-extorter	5–72 h
Extorter	Find-victims	Shops available	Shop	Wait-for-extorter	72 h
Extorter	Find-victims	No shops available	Shop	Become-active	150–200 h
Extorter	Give-up	Banished by a more successful extorter	Extorter	Become-active	24–72 h
Extorter	Leave-jail		Extorter	Become-active	12–36 h
Police	Start-prosecute	Started	Police	Put-to-custody	0–5000 h
Police	Put-to-custody	Escaped	Extorter	Become-active	360–720 h
Police	Put-to-custody	Not escaped	Police	Convict	720–5000 h
Shop	Wait-for-extorter	Denounced	Police	Start-prosecute	12–84 h
Shop	Wait-for-extorter	Denounced	Extorter	Punish	0–168 h
Shop	Wait-for-extorter	Denounced	Extorter	Become-active	24–168 h
Shop	Wait-for-extorter	Paid	Extorter	Return	600–700 h
Shop	Wait-for-extorter	Neither denounced nor paid	Extorter	Punish	0–168 h
Shop	Wait-for-extorter	In criminal records	Police	Start-prosecute	12–83 h
Shop	Wait-for-extorter	Not met	Extorter	Return	600–700 h
Shop	Decide-to-pay-or-not	Nothing paid	Extorter	Become-active	8–12 h
Shop	Decide-to-pay-or-not	Paid	Extorter	Become-active	160–240 h
Shop	Decide-to-pay-or-not	Refused to pay	Extorter	Become-active	8–12 h

the actions listed in Table 8.2 there are norm invocations (which happen immediately, as in the period-oriented version) and periodic events which happen:

- Once a week
- Consumers go shopping or
- Once a month
- Shops pay salaries to consumers in the latter's role of employees
- The state (which is represented by the NetLogo observer, as is the court) compensates extortion victims from confiscated assets if there are any
- Extorters pay into a funds from which needy extorters of the same family are subsidised
- Statistics are collected

To describe how the model works in detail it will be best to list the actions which are taken in a sequence by the respective agents.

The initialisation puts agents of all four types to patches in a way that no patch is occupied by more than one agent. Agents stay on their patches throughout a simulation run; that is, they do not move for their actions (one could also say that they immediately return to their patches after they have performed an action which in reality necessitates a visit at a distant place, for instance—for an extorter—to take the extortion money home from a victim or—for a police officer—to arrest an extorter or—for a customer—to buy something from a shop or to work for a shop).

Immediately after the initialisation all consumers are scheduled to go shopping at 10:00 of the first day (and to repeat this action once a week at the same time) and to select a shop from whom they will buy—later they will have an opportunity to switch to another shop when their norm salience calculations recommend them to buy from a shop which does not pay extortion. At the same time the extorters are scheduled to become active after a delay of 3–5 days. Moreover the periodic events mentioned above are scheduled.

Once an extorter becomes active it starts to find possible victims in its vicinity (whose initial radius is given by a parameter valid for all extorters but which can be extended by a factor which is another input parameter whenever the search for victims turns out unsuccessful). If a victim was found it is approached after a small delay (the time between find-victims and wait-for-extorter in Table 8.2); if not, another attempt at finding victims is made with a delay between approximately 6 and 8 days (150 and 200 h).

If victim and extorter meet, the former makes a decision whether it denounces the latter or whether it pays the requested amount to the latter—the requested amount is a percentage of the current income from sales determined by an input parameter. If the shop decides not to pay or if it has not had any income during the current month it cannot pay, the extorter is scheduled to return within a week. In these cases two outcomes are possible:

If the shop refused to pay in the first meeting the extorter will only be successful if the shop reconsiders its decision neither to denounce nor to pay or in case the shop was unable to pay anything before the extorter might be successful if the shop in turn was meanwhile successful in selling anything to a paying customer; otherwise the shop will be bankrupt and perhaps be reopened when extortion and/or punishment was the reason of its bankruptcy and the state compensated it from confiscated extorter assets.

If the victim decides to denounce it asks a police officer nearby and schedules its start-prosecution procedure for some time within the next 2 weeks, and there is also a chance that the police officer has observed the extorter's approach (but only if the latter is already in the police's criminal records, i.e. was denounced earlier on by some other shop or who was found punishing another shop—an activity which is always observable and cannot be concealed, much like arson), and in this case the prosecution also starts within a week. After the extorter was denounced it is scheduled to become active again to find other victims during the following 2 weeks (but if it is meanwhile caught by the police the scheduled task will, of course, never be performed). As a further consequence of denunciation, the denounced extorter will also plan to punish the shop, and this is scheduled for some time within the next 2 weeks, provided that this extorter has not been brought to custody before this date.

If the same victim is approached by several extorters before the former makes the decision between denouncing and paying discussed above, it has to choose among the competing extorters. The successful extorter subordinates its competitors, thus forming a growing hierarchy of families which is documented in one of the NetLogo windows (but not analysed in depth so far). The unsuccessful extorters will become active again and try to find victims during the next few days. At the end of each month the extorters' incomes are redistributed within each family (isolated extorters do not participate in this redistribution process).

When a police officer starts to prosecute an extorter (either after denunciation or after police observation) it will take up to 200 days until the extorter is either brought to custody or escaped. In the latter case the extorter will become active again; otherwise it stays in custody until the court (represented by NetLogo's observer) passes a sentence, which will take between 1 and about 7 months. If the extorter is acquitted it becomes active the next day, and if it is sentenced his or her period of being inactive in jail will be an integer number of years (between three and eight). When the prisoner is released it will again become active and try to find victims within the next few days.

8.3 Input Parameters and Output Metrics

Model runs are initialised with a number of input parameters, not all of which will turn out relevant in the end (and, indeed, sensitivity analysis of earlier versions of the NetLogo model showed that many are not), but all of them were kept for the event-oriented version to find out whether they might become relevant under the new circumstances. The parameter space is spanned by uniform distributions of the global variables listed in Table 8.3. The simulation model was run 1280 times

Table 8.3 List of input parameters for the sensitivity analysis

Name	Function	Minimum value	Maximum value
Background (bg)	Fills the memories of all agents with observations and invocation referring to their respective norms	–5	5
NDW (ndw)	Weight of the normative drive; the weight if the individual drive is 1–NDW	0.3	0.7
Discount (disc)	Multiplier applied to each counter whenever new norm invocations arrive	0.8	1.0
Local (loc)	Distance within which norm-relevant observations can be made and norm invocations can be received	2	10
Benefit-for-victims (bfv)	Value of the protection offered by an extorter for the extortion money requested as a percentage of the requested sum	75	150
Conviction-probability (cvp)	Probability that the court will sentence an arrested extorter for a longer period in jail	0.1	0.8
Extortion-level-low[a]	Lower bound of the percentage of the current turnover from sales of a shop requested by an extorter, the upper bound being fixed as 10	5	10
Punishment-severity-low	Lower bound of the percentage of the current wealth of a shop requested robbed by an extorter after refusal to pay, the upper bound being fixed as 10	10	10
Escape-chance (ec)	Probability that an extorter is not caught by the prosecuting police	0.1	0.8
Extortion-radius-extension	Multiplier of the initial distance within which an extorter can find victims, applied after each unsuccessful search	1.0	1.2
Vision-range (vr)	Distance within which a shop can find a police officer to whom it can report an extortion attempt	5	85
Hide-denounce-propensity	Probability that a shop publishes its readiness not to denounce an extortion attempt	0	0.125

The abbreviations in the first column of the table are used in Table 8.5. Input parameters without such an abbreviation do not have any significant influence in the linear regressions reported in Table 8.5.

[a]Extorters can individually differ in their extortion level and punishment severity instance variables; hence in principle four types of extorters are possible: those with both levels low, those with both levels high and those with low (high) extortion level and high (low) punishment severity. This typology stems from earlier versions and did never turn out important, and with the values reported in Table 8.3 only two types occur as the severity level is constant

Table 8.4 Output metrics of the simulation

<table>
<tr><td>Percentage of undetected cases</td><td rowspan="5">All of these measured as moving averages over the recent two simulated months, percentages are based on all attempts which happened within the respective 2 months, including those which only became known to extorter and victim, i.e. there is no "dark figure"</td></tr>
<tr><td>Percentage of undenounced extortions</td></tr>
<tr><td>Percentage of completed extortions</td></tr>
<tr><td>Percentage of cases with arrest</td></tr>
<tr><td>Percentage of cases with punishment</td></tr>
<tr><td>Percentage of critical consumers</td><td>Measured as the moving average of the means of the individual propensity to buy only from non-denouncing shops</td></tr>
<tr><td>Denunciation rate</td><td rowspan="5">All of these measured as moving averages over the recent two simulated months, percentages are based on all attempts which happened within the respective 2 months, not including those which only became known to extorter and victim, i.e. these rates are better comparable to empirical data</td></tr>
<tr><td>Prosecution rate</td></tr>
<tr><td>Success rate</td></tr>
<tr><td>Arrest rate</td></tr>
<tr><td>Conviction rate</td></tr>
</table>

with 100 shop agents, 50 extorter agents, 10 police agents and 800 consumer agents; each run lasted two simulated years.[3]

The list of output metrics which can be used for statistical analyses is even longer. It can be found in Table 8.3.

Besides this list, the graphical user interface offers a variety of plots, both showing the history of some of the metrics listed in Table 8.4 and some histograms of norm saliences and action propensities changing over time. The history of some of the output metrics is stored in separate files for each run, and the distribution of norm saliences at the end of the run is also available for statistical analysis.

8.4 Results of 1280 Simulation Runs

The results of the event-oriented version of the model are fairly similar to the results of the period-oriented version, as Fig. 8.1 shows. Not surprisingly there is a high correlation of the 2 % of undenounced and of completed extortions, as extortions are either successful from the point of view of the extorter or denounced—except the case when the victim is unable to pay at all but does not want to denounce. The correlations between these two output metrics and the input parameters of the model are smaller than in the period-oriented case, presumably due to the higher

[3] Runs with different numbers of agents per type—e.g. three times as many as mentioned in footnote 16—generate results which are very similar to the ones reported here, but consume disproportionate computing time.

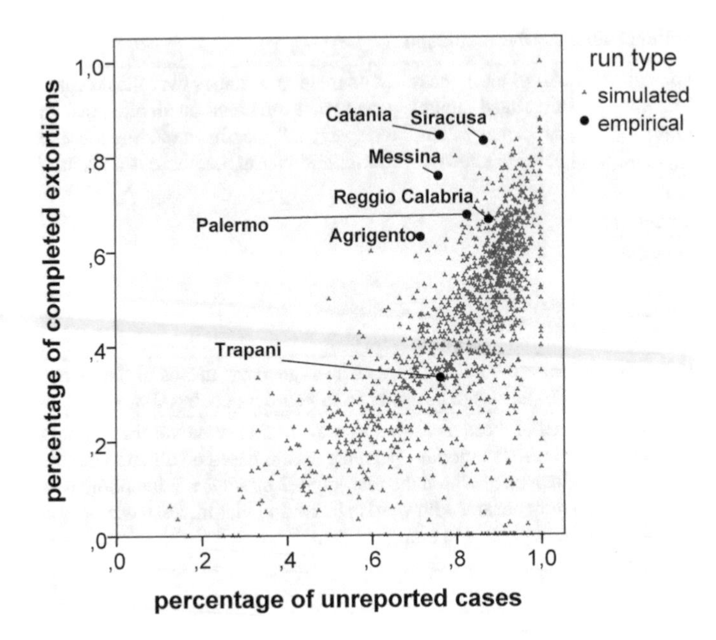

Fig. 8.1 Scattergram of the two main output metrics: percentage of undenounced cases vs. percentage of completed extortions

Table 8.5 Regression coefficients and reduction of the variance of the main output metrics by input parameters

Per cent of all cases	R^2	Standardised regression coefficients							
		Bg	Loc	disc	ndw	bfv	cvp	vr	Ec
Undetected	.312	−.545	−.055	.102	n.s.	.064	n.s.	n.s.	n.s.
Unreported	.424	−.623	n.s.	n.s.	−.123	.077	−.076	n.s.	.062
Completed	.308	−.460	n.s.	n.s.	.267	.095	n.s.	.124	n.s.
Arrest	.118	.091	n.s.	−.072	.124	−.061	.109	.139	−.241
Conviction	.325	.522	n.s.	−.046	−.217	−.091	n.s.	n.s.	n.s.
Per cent of detected cases	R^2	*bg*	*Loc*	*disc*	*ndw*	*bfv*	*cvp*	*vr*	*Ec*
Denounced	.357	.585	n.s.	n.s.	n.s.	−.096	.053	−.073	n.s.
Successful	.290	−.378	n.s.	n.s.	.341	.104	−.067	.136	n.s.
Prosecuted	.368	−.100	.074	−.085	.474	n.s.	−.062	.335	.067
Arrested	.119	−.055	.054	n.s.	.144	n.s.	.109	.146	−.249
Convicted	.028	n.s.	n.s.	n.s.	n.s.	n.s.	.140	n.s.	−.062

randomness which comes into the model as a consequence of the random delays between triggering action and triggered action which leads to path dependence: if one runs the model with identical input parameters but different seeds of the random number generator, then for the first about 150 days the difference between runs is small, but from then the random events cumulate and increase the variance between

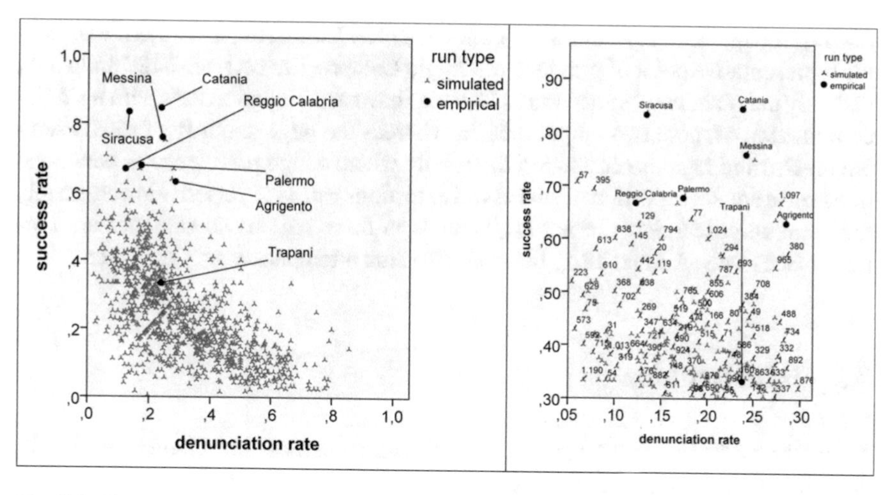

Fig. 8.2 Comparison between provinces in Southern Italy and simulation runs: cases which came into the open

runs considerably. There are significant influences of a number of input parameters (see Table 8.5), but the overall reduction of variance is only about 30 %, and although the contributions of the third and fourth less important input parameters are often smaller than 1 %, they are still significant on a 5 % level, which is only due to the large number of simulation runs, another hint at the fact that the effect is more meaningful than the level of statistical significance (cf. Ziliak & McCloskey, 2007).

As in the case of the period-oriented version, most of the empirical cases—calculated from the Sicily and Calabria database for seven provinces (Frazzica et al., 2015)—are situated in the margin of the frequency distribution of the simulation runs, except for the provinces of Trapani[4] and Reggio Calabria. The main reason for this weak validation of the model is that the empirical data are blurred with all those cases which never became known to the police, the prosecutors or the media and thus remained in the dark—which cannot happen in a simulation where all attempts at extorting are documented in the simulation output. As the simulation model also reports the extortions which were only known to extorter and victim these can be used as an estimate for "dark cases", and denounced and completed extortions can be calculated as percentages of all the simulated cases which came into the open (denounced or observed by the police without the help of the victim). If one uses these percentages instead of the raw percentages as in Fig. 8.2, the validation is more successful, and simulation runs matching the empirical data of some of the provinces can be identified.

Still, most of the provinces can be found at the margin of the joint distribution of the two output metrics comparable to empirical data, but a few simulation runs can be identified whose results are similar to the provinces of Trapani (#990 and many

[4] That the province of Trapani is different from the other provinces was already clear from the findings in Chap. 9.

others), Agrigento (#1097 and #521), Messina (#342 and #355) and Palermo (#181); the representative point of province of Reggio Calabria lies between #129 and #395; and only the other two (Siracusa and Catania) have no simulation runs with a similar combination of these two output metrics whereas the large majority of runs resemble much more regions outside Southern Italy where a high prevalence of denunciations and a small proportion of successful extortions can be expected—unfortunately for these empirical data are not available as in these regions no statistics are produced (see Sects. 9.3 and 12.1) because extortion attempts are very rare there.

8.5 Interventions

As in the period-oriented version, interventions are possible (Troitzsch, 2016a, 2016b). At a certain point of time, e.g. at the end of the first year, all memories of all agents of one or several or all types are emptied and refilled with a number of invocations of "liberal" norms (the norms listed in the first row of each pair of rows in Table 8.1). In single runs the effect of such an intervention can immediately be observed via the GUI; in batch runs a CSV file is written which can be analysed afterwards.

Figure 8.3 shows the two main output metrics over two simulated years with all 16 possible interventions applied at the end of the first simulated year (the combinations of interventions are partly suppressed in these two diagrams showing only dashed curves to enable readers to follow the histories of the four runs in which the agent of only one agent type was subject to intervention). During the first simulated year all runs have an identical history as all of them were started with the same seed of the pseudorandom number generator. Immediately after the intervention injecting a high number of liberal or civic norm invocations the histories run apart: when the intervention is only applied to the shop agents, very soon all extortion attempts are denounced, and none is successful. Applying the interventions to the agents of any of the other types has only a moderate effect, which is, however, still better than no intervention at all. To find out whether combinations of interventions to the agents of more than one type have special effects, the diagram in Fig. 8.4 shows the values of the two main output metrics about six simulated weeks after the intervention at the end of the first year (to be precise: on day 400) for 10 groups of runs (different pseudorandom number generator seeds between groups, different intervention targets within each group). One sees that

The distribution of the outcomes of interventions only to shops does not differ from the distribution of the outcomes of intervention applied to the agents of all four types or the distribution of the outcomes of intervention applied to shops and any other agent type.

The distribution of the outcomes of intervention to all but the shops (mean 0.6591) does not significantly differ from the distributions produced when no intervention was applied at all.

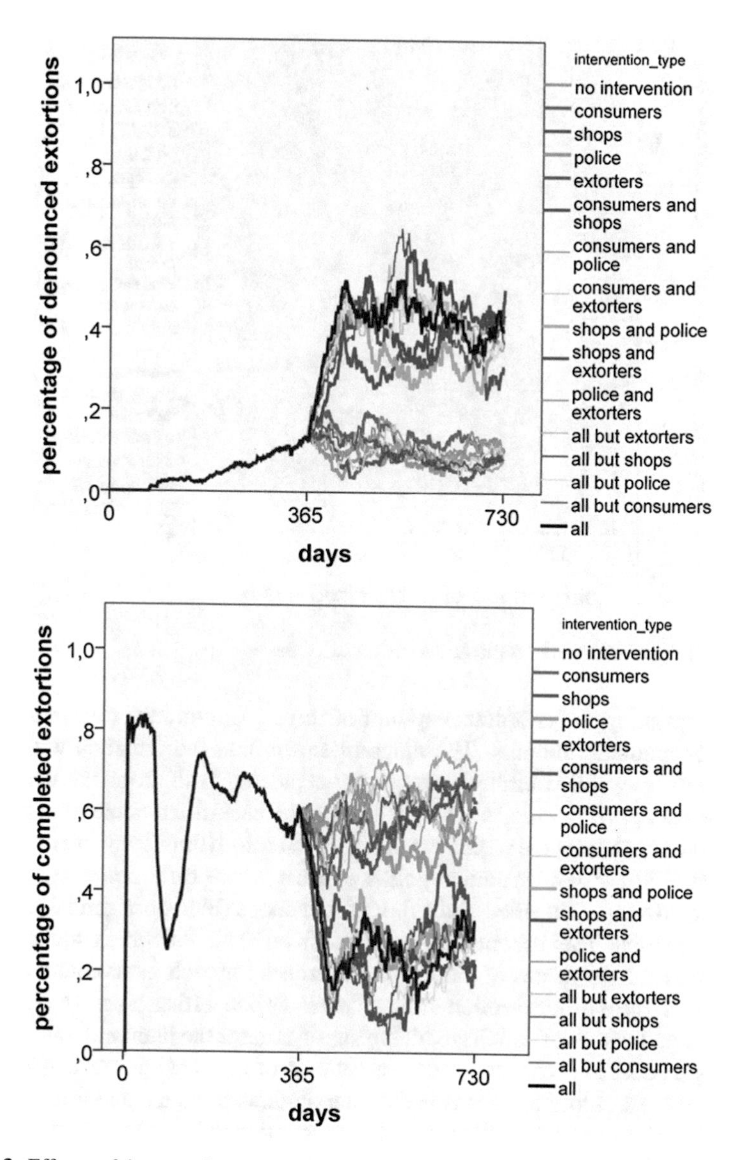

Fig. 8.3 Effects of interventions on the percentage of denounced and completed extortions, respectively

For more details of these results see Table 8.6 which shows that for nearly all output metrics it is only the interventions including shops that differ significantly—and with absolute t-values beyond 8—from the no-intervention cases. Only the output metrics referring to arrest and conviction differ from these overall diagnosis. Figure 8.4 shows the state of these 10 groups of 16 runs each with the 16 possible intervention combinations 12 simulated months after the interventions; again the 2

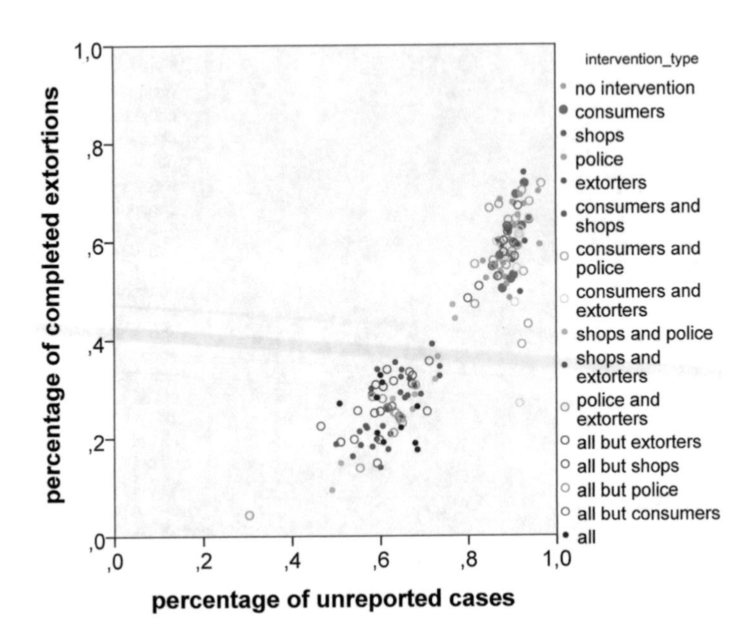

Fig. 8.4 Effects of interventions on the success of extortion attempts

variables spanning the coordinate system of these scattergrams are averages over the past 2 simulated months. The diagram shows that the ten runs without any intervention (grey-filled circles) are strictly separated both from the ten runs in which only shops (blue-filled circles) or all agents (black-filled circles) were subject to intervention whereas the region occupied by the runs without any interventions is also filled with the representative points of runs where only police (green-filled circles) or extorters (red-filled circles) or all but shops (blue open circles) had their memories refilled. This becomes even clearer from Table 8.6 which shows the difference between the means of the moving averages for each intervention type and the mean of the no-intervention runs as well as the effect sizes (measured as Dunnett's t and η^2), all of which are highly significant for the interventions including the shops ($\alpha < 0.0005$) except for the percentage of cases with arrest. And for all extortion-related output metrics it is the interventions in which shops are involved which differ most from the no-intervention runs. Interventions directed to agents of the other three types (and only to these) have a very small effect, and interventions directed to two or three agent types leaving out the shops have only an intermediate effect which is not statistically significant. It seems that the chance of getting hold of criminals does not depend on any interventions—there are some differences, but they are not significant.

The output metric "percentage of critical consumers" follows another pattern. Its η^2 is fairly high (and significant) with 0.398, but the effect in mean differences is only modest: without intervention the value of this output metric is 21.7 %; with interventions applied to consumers it increases to 29.8 %; if other groups are also

Table 8.6 Effects of interventions: Moving averages over two simulated months of output metrics one simulated year after the intervention

Comparisons between intervention with … addressed and no intervention	Standard error	η^2		Con-sumers	Shops	Police	Extorters	Consumers and shops	Consumers and police	Consumers and extorters	Shops and police	Shops and extorters	Police and extorters	All but extorters	All but shops	All but police	All but consumers	All
Percentage of undetected cases	0.0611	0.631	MD	−0.027	−0.288	0.046	−0.013	−0.346	−0.118	−0.006	−0.364	−0.329	−0.006	−0.338	−0.053	−0.313	−0.388	−0.433
			Dt	−0.437	**−4.718**	0.748	−0.213	**−5.657**	−1.932	−0.097	**−5.957**	**−5.383**	−0.105	**−5.525**	−0.860	**−5.126**	**−6.345**	**−7.085**
Percentage of unreported cases	0.0593	0.731	MD	−0.054	−0.399	−0.047	−0.015	−0.498	−0.030	−0.040	−0.429	−0.471	−0.052	−0.463	−0.069	−0.448	−0.350	−0.512
			Dt	−0.919	**−6.736**	−0.801	−0.260	**−8.401**	−0.513	−0.682	**−7.237**	**−7.941**	−0.876	**−7.811**	**−1.166**	**−7.554**	**−5.898**	**−8.643**
Percentage of denounced extortions	0.0593	0.731	MD	0.054	0.399	0.047	0.015	0.498	0.030	0.040	0.429	0.471	0.052	0.463	0.069	0.448	0.350	0.512
			Dt	0.919	**6.736**	0.801	0.260	**8.401**	0.513	0.682	**7.237**	**7.941**	0.876	**7.811**	1.166	**7.554**	**5.898**	**8.643**
Percentage of completed extortions	0.0675	0.682	MD	−0.017	−0.377	0.127	0.079	−0.407	−0.053	0.040	−0.331	−0.348	0.000	−0.360	0.044	−0.391	−0.417	−0.428
			Dt	−0.248	**−5.586**	1.880	1.176	**−6.023**	−0.786	0.599	**−4.904**	**−5.148**	0.002	**−5.335**	0.645	**−5.783**	**−6.180**	−6.340
Percentage of cases with arrest	0.0342	0.151	MD	0.022	0.029	0.087	0.034	0.023	0.058	0.033	0.103	0.067	0.076	0.036	0.024	0.057	−0.005	0.084
			Dt	0.656	0.858	2.548	0.998	0.676	1.680	0.957	3.018	1.957	2.222	1.049	0.687	1.651	−0.138	2.453
Percentage of cases with punishment	0.0652	0.606	MD	0.030	0.261	−0.074	−0.038	0.349	0.072	−0.003	0.293	0.311	−0.026	0.307	−0.002	0.328	0.400	0.373
			Dt	0.460	**3.999**	−1.138	−0.589	**5.355**	1.108	−0.040	**4.489**	**4.761**	−0.394	**4.711**	−0.032	**5.026**	**6.135**	**5.714**
Denunciation rate	0.0757	0.663	MD	0.083	0.485	0.128	−0.010	0.553	−0.001	0.056	0.446	0.535	0.047	0.516	0.098	0.511	0.342	0.507
			Dt	1.099	**6.409**	1.692	−0.129	**7.295**	−0.015	0.743	**5.890**	**7.059**	0.615	**6.818**	1.290	**6.744**	**4.516**	**6.694**
Prosecution rate	0.0840	0.305	MD	0.067	−0.076	0.260	0.187	−0.145	0.029	0.063	−0.005	−0.124	0.151	0.057	0.129	−0.183	0.019	−0.038
			Dt	0.792	−0.909	3.098	2.222	−1.732	0.350	0.750	−0.063	−1.475	1.795	0.684	1.540	−2.176	0.225	−0.450
Success rate	0.0736	0.460	MD	−0.124	−0.289	0.050	0.091	−0.306	−0.042	−0.010	−0.258	−0.299	−0.197	−0.286	−0.021	−0.312	−0.288	−0.298
			Dt	−1.683	**−3.928**	0.680	1.235	**−4.156**	−0.574	−0.136	**−3.504**	**−4.065**	**−2.674**	**−3.889**	−0.282	**−4.236**	**−3.921**	**−4.052**
Arrest rate	0.0791	0.204	MD	0.039	0.009	0.253	0.072	0.009	0.108	0.066	0.094	0.062	0.293	0.012	0.037	0.062	−0.039	0.061
			Dt	0.492	0.108	**3.202**	0.908	0.116	1.361	0.836	1.194	0.781	**3.708**	0.157	0.465	0.778	−0.495	0.777
Conviction rate	0.0070	0.076	MD	−0.011	−0.011	−0.011	−0.011	−0.011	−0.003	−0.002	−0.001	−0.011	−0.011	−0.006	−0.011	−0.011	−0.011	−0.005
			Dt	−1.591	−1.591	−1.591	−1.591	−1.591	−0.398	−0.289	−0.098	−1.591	−1.591	−0.795	−1.591	−1.591	−1.591	−0.749

Values relate to groups of 10 runs of each intervention types, all runs of a group are started with the same pseudorandom number generator, “standard error” is the standard error of the distribution of the variables for the no-intervention runs, “MD” is the mean deviation between the no-intervention runs and the respective intervention runs, “Dt” is Dunnett’s two-sided t; t-values >2 and <−2 are in *bold* involved the mean over all ten runs increases to 29.9 %—the high η^2 is mainly due to the fact that the variance within the groups of size ten is very small, and the t-value is not even significant with 1.01.

This is certainly in line with the experience of high police officers in Sicily, one of which recently—after a success in getting hold of a number of high-ranked Mafiosi as a consequence of a series of denunciations made by entrepreneurs—was cited by a Sicilian newspaper with the remark that denunciation is worth the risk and useful (La Stampa, 03/11/2015), and it is also in line with the strategy of the Addiopizzo Movement (Vaccaro & Palazzo, 2015) although it is also argued that the revolution was on the side of the consumers, as the title of a recent book (Di Trapani & Vaccaro, 2014, 2016) announced that it was the attempt at leveraging consumers' responsible purchase that fought the mafia, but the authors also admit that the movement began with a list of "pizzo-free" entrepreneurs (and the Italian blurb says that the book is "a homage to all those who took the personal risk to affirm the values of legality and liberty"), and hence even they will accept that convincing entrepreneurs that denunciations are worth the risk was the main cause of Addiopizzo's success.

8.6 Summary

When one compares the event-oriented NetLogo version of the GLODERS model to its own period-oriented predecessors (with or without the agents' norm orientation) they have one feature in common: The joint distribution of the two main output metrics is more or less the same in all three versions, and all three versions predict the empirical cases of Southern Italy only at the margin of this distribution as most Monte Carlo runs show a behaviour which one would expect from regions where extortion is rare, rarely successful and often denounced. Hence neither the introduction of norm-oriented agent behaviour nor the introduction of an event-oriented action scheduling changed the overall behaviour of the model. GLODERS-S was purposefully calibrated to match the provinces in Southern Italy and was successful at least with respect to the province of Palermo and the scenarios between 1980 and 2015 (see Fig. 7 in Nardin et al. (2016)).

On the other hand there are a number of differences between the three NetLogo versions. The step from a simple stochastic model where agents made their action decisions only based on constant action probabilities to the period-oriented model with agents influencing each other and making their decision based on calculations of norm saliences introduced an additional complexity which led to more complicated trajectories of the output metrics which showed traces of path dependencies, among others, with the result that final outcomes depended much more on early events and that the proportion of the variance of output metrics explained by input parameters was reduced. The event orientation with its stochastically defined delays between triggering action and triggered action allowed for even more path dependence and less variance reduction which could best be seen in the variance of the 20 runs per intervention type in Fig. 8.4 where all of these runs were determined by the exactly equal set of input parameters and differed only in the seed of the pseudorandom number generator.

Further research and an even deeper comparison between the extortion racket simulation models will show whether it was worthwhile to deviate from the KISS ("keep it simple, stupid!") principle which was followed in the simple stochastic version (Troitzsch, 2015a, 2015b) and to expand the models according to the more descriptive KIDS (Edmonds & Moss, 2005) version where the actions are taken by the agents in a more sophisticated way and where the periods between actions taken by the agents of different kinds are explicitly modelled (albeit without much empirical background).

References

Di Trapani, P., & Vaccaro, A. (2014). *Addiopizzo. La rivoluzione dei consumi contro la mafia*. Cagliari: Arkadia.

Di Trapani, P., & Vaccaro, A. (2016). *Addiopizzo: Leveraging consumers responsible purchase to fight mafia*. Madrid: McGraw-Hill Interamericana de España.

Edmonds, B., & Moss, S. (2005). From KISS to KIDS—An 'anti-simplistic' modelling approach. In P. Davidsson, B. Logan, & K. Takadama (Eds.), *Multi-agent and multi-agent-based simulation: Joint Workshop MABS 2004* (S. 130–144). Berlin: Springer.

Frazzica, G., Punzo, V., La Spina, A., Militello, V., Scaglione, A., & Troitzsch, K. G. (2015). *Sicily and calabria extortion database*. (GESIS, Hrsg.) Von Datorium: http://dx.doi.org/10.7802/1116 abgerufen.

Nardin, L. G., Andrighetto, G., Conte, R., Székely, Á., Anzola, D., Elsenbroich, C., et al. (2016). Simulating the dynamics of extortion racket systems: A sicilian Mafia case study. *Autonomous Agents and Multi-Agent Systems*

Sheppard, C., & Railsback, S. (2014, August 28). *NetLogo time extension*. Retrieved February 18, 2016, from https://github.com/colinsheppard/time.

Tisue, S., & Wilensky, U. (2004). NetLogo: Design and implementation of a multi-agent modeling environment. *Agent2004 Conference*. Chicago

Troitzsch, K. G. (2015a). Extortion racket systems as targets for agent-based simulation models. Comparing competing simulation models and Emprical data. *Advances in Complex Systems, 18*, 1550014.

Troitzsch, K. G. (2015b). Distribution effects of extortion racket systems. In F. M. Amblard (Ed.), *Advances in artificial economics* (pp. 181–193). Berlin: Springer.

Troitzsch, K. G. (2016). *Using empirical data for designing, calibrating and validating simulation models*.

Troitzsch, K. G. (2016). Can agent-based simulation models replicate organised crime? *Trends in Organised Crime*

Vaccaro, A., & Palazzo, G. (2015). Values against violence: Institutional change in societies dominated by organized crime. *Academy of Managemtn Journal, 58*(4), 1075–1101.

Ziliak, S. T., & McCloskey, D. N. (2007). *The cult of statistical significance. How the standard error costs us jobs, justice and lives*. Ann Arbor: University of Michigan Press.

Chapter 9
Survey Data and Computational Qualitative Analysis

Klaus G. Troitzsch

9.1 Social Norms as Objects of Transnational Surveys

The available survey data sources can only give an incomplete access to the empirical problem under consideration for the reasons discussed in more depth in Sect. 12.1, but anyway three surveys can be used as sources for finding out which norm orientations prevail in the regions under consideration in GLODERS and which are the propensities and frequencies of norm-related actions.

To determine the initialisation of individual and normative weights for simulations of extortion racket systems, it is desirable to have some empirical information on the distribution of related traits in the populations of the regions which deliver scenarios for the design of the simulation model. The most easily usable source of information on this topic is the European Values Study (EVS, 2008) the data of which were documented on a very detailed regional level (down to NUTS-3, but for the ERS purposes NUTS-1 will be sufficient).

The questions asked in the EVS are not exactly aimed at finding out the interviewees' orientation towards norms, but the set of questions listed in Table 9.1 can be used as a surrogate.

Note that these variables are coded with 1 for mentioned, and 2 for not mentioned.

A factor analysis with varimax rotation yields three factors (Table 9.2), two of which seem to be at least loosely related to the individual (IW) and normative (NW) weights used in the agent's decision process.

Factor 1 can be interpreted as a trait of people who want their children to be able to make their decisions independently and according to their feeling of being responsible for their actions and not just according to what they are told to do by others or

K.G. Troitzsch (✉)
Computer Science Department, Universität Koblenz-Landau,
Universitätsstraße 1, Koblenz, Rheinland-Pfalz 56070, Germany
e-mail: kgt@uni-koblenz.de

C. Elsenbroich et al. (eds.), *Social Dimensions of Organised Crime*,
Computational Social Sciences, DOI 10.1007/978-3-319-45169-5_9

Table 9.1 Values which children should learn

Q52: Here is a list of qualities which children can be encouraged to learn at home. Which, if any, do you consider to be especially important? Please choose up to five!		
v170	A	Good manners
v171	B	Independence
v172	C	Hard work
v173	D	Feeling of responsibility
v174	E	Imagination
v175	F	Tolerance and respect for other people
v176	G	Thrift, saving money and things
v177	H	Determination, perseverance
v178	I	Religious faith
v179	J	Unselfishness
v180	K	Obedience

Table 9.2 Factor analysis of values which children should learn (varimax rotated component matrix)

	Component		
	1	2	3
v173 Learn children at home: feeling of responsibility (Q52D)	−.640		
v180 Learn children at home: obedience (Q52K)	.544	−.277	
v178 Learn children at home: religious faith (Q52I)	.527		
v170 Learn children at home: good manners (Q52A)		−.557	
v174 Learn children at home: imagination (Q52E)		.520	
v177 Learn children at home: determination, perseverance (Q52H)		.495	
v171 Learn children at home: independence (Q52B)	−.404	.487	
v172 Learn children at home: hard work (Q52C)			−.690
v175 Learn children at home: tolerance+respect (Q52F)		−.365	.520
v179 Learn children at home: unselfishness (Q52J)	.416		.476
v176 Learn children at home: thrift (Q52G)			−.474
Extraction method: principal component analysis			
Rotation method: varimax with Kaiser normalization			

by religion; persons with high positive scores in this factor will have a high individual weight, and those with high negative scores will have a high normative weight.

The interpretation of factor 2 is in a way similar; it is a trait of people who want their children to have good manners and who are less interested in their independence and perseverance; persons with high positive scores in this factor will have a high normative weight.

Both factors are, according to the analysis method applied, uncorrelated; thus it might be difficult to decide which of the two is a better surrogate to the normative and individual weights.

Another way to find one dimension (except using only one factor from factor analysis) is a one-dimensional optimal scaling. The overall correlation of this scale with the two factors mentioned above is –0.595 and –0.747, respectively. Thus perhaps this

Table 9.3 Relation between factors and important items

v173 Learn children at home: feeling of responsibility (Q52D)		N	Mean	Std. deviation	Std. error mean
Factor 1	1 mentioned	43,759	0.40	0.76	0.0036
$t=204.759$	2 not mentioned	16,701	−1.04	0.79	0.0061
Factor 2	1 mentioned	43,759	0.09	0.97	0.0047
$t=34.363$	2 not mentioned	16,701	−0.23	1.03	0.0080
Optimal scale	1 mentioned	48,012	0.20	0.96	0.0044
$t=89.310$	2 not mentioned	17,101	−0.56	0.98	0.0075
v180 Learn children at home: obedience (Q52K)		N	Mean	Std. deviation	Std. error mean
Factor 1	1 mentioned	16,583	−0.088	0.84	0.0065
$t=-159.377$	2 not mentioned	43,877	0.33	0.84	0.0040
Factor 2	1 mentioned	16,583	0.45	0.89	0.0069
$t=74.085$	2 not mentioned	43,877	−0.17	0.98	0.0047
Optimal scale	1 mentioned	18,681	−0.97	0.85	0.0062
$t=-182.099$	2 not mentioned	44,552	0.37	0.85	0.0040

is a variable whose distribution in the scenario regions could be used for initialising simulation runs.

To clarify the direction of the newly created scales, Table 9.3 shows their mean values for the two groups of respondents who mentioned and did not mention, respectively, “obedience” and “feeling of responsibility” among their five important “qualities which children can be encouraged to learn at home”.

The result is that all newly created scales are positively correlated with the importance of children learning a feeling of responsibility as those who mentioned this have a positive mean score in all three scales whereas those who did not mention this item as important have negative mean scores. For the obedience item this is a little less clear, although for factor 1 and the dimension created by one-dimensional optimal scaling those who mentioned the item as important have a high negative mean score, but this does not hold for the second factor. One must also observe that a large proportion of the respondents seem to have characterised both items as important.

Taking into account the combinations of the respondents’ opinions with respect to these two items yields Table 9.4.

All differences are significant, but this is mostly due to the very large sample. If one only compares the means for those who mentioned exactly one of the two items as important, the t statistics for factor 1 is 256.075, for factor 2 it is −29.155, and for the dimension created by the one-dimensional optimal scaling it is 176.267. Thus factor one or the one-dimensional score should be preferred for further analyses.

The distribution of all three scores derived in the past subsection in the regions of interest is shown in Table 9.5.

The differences between the means of the regions are highly significant. If one excludes the “Other” regions from the analysis and restricts the analysis to a comparison of the four regions of our interest, the etas are 0.304, 0.149, and 0.336, respectively (which is not extremely high as the variance within the groups is nearly

Table 9.4 Relation between the two important items and the factors

Report				
v173_v180 Importance of obedience and responsibility		FAC1_1 REGR Factor score 1 for analysis 1	FAC2_1 REGR Factor score 2 for analysis 1	OBSCO1_1 Object scores dimension 1
11 Both items important	Mean	−.4457385	.5441746	−.7116
	N	10,271	10,271	11,459
	Std. deviation	.60654804	.90009265	.73091
12 Only responsibility important	Mean	.6532856	−.0522376	.5271
	N	33,488	33,488	33,949
	Std. deviation	.59856365	.95164250	.79186
21 Only obedience important	Mean	−1.5990032	.2993194	−1.2755
	N	6313	6313	6443
	Std. deviation	.64869065	.86437789	.74487
22 Both items unimportant	Mean	−.6935965	−.5514676	−.1204
	N	10,389	10,389	10,545
	Std. deviation	.66232943	.99280386	.83262
Total	Mean	.0000000	.0000000	.0040
	N	60,461	60,461	62,397
	Std. deviation	1.00000000	1.00000000	1.01405

as high as in the whole sample, but the differences between the four means are high enough). Thus one would initialise the normative drive weight (NDW) of Southern Italy with a distribution whose mean is greater than 0.5, the mean of Northern Italy (if it is used at all in later scenarios) would be approximately 0.5, whereas the Netherlands and particularly South-West Germany would start with a distribution whose mean is below 0.5.

Converting the scales discussed above (which are distributed around a mean of approximately 0 and range from about −3 to +3) into normative weights ranging from 0 to 1 is not straightforward as it is entirely unclear whether the individual and normative weights which one would find in the (fictitious) average European are 0.5 or any other number in the interval between 0 and 1. But it seems reasonable to start with 0.5 as the weight for the fictitious average European and then to convert the scale with a sigmoid function. This would lead to distributions in the four regions as shown in Fig. 9.1 (where the conversion function is the arcus tangens, to be exact: $N = \left(\arctan\left(s\right) + \pi / 2\right) / \pi$). This yields estimates for the input parameter normative drive weight (NDW) of 0.31, 0.42, 0.51, and 0.59, respectively (see Fig. 9.1).

Although the scores originating from factor analysis and from one-dimensional optimal scaling convey more information and less noise than the original observable items, it is interesting enough to analyse the two items selected above to support the

Table 9.5 Factor scores in GLODERS regions

GLODERS regions		Factor 1	Factor 2	Optimal scale
Baden-Württemberg	Mean	0.75	−0.27	0.89
	N	168	168	169
	Std. deviation	0.74	1.05	0.96
The Netherlands	Mean	0.30	0.24	0.34
	N	1526	1526	1552
	Std. deviation	0.88	1.00	0.98
Northern Italy	Mean	−0.73	0.19	−0.03
	N	697	697	706
	Std. deviation	0.93	0.97	0.98
Southern Italy and Islands	Mean	−0.30	0.42	−0.45
	N	525	525	533
	Std. deviation	0.90	0.87	0.98
Other	Mean	−0.01	−0.01	−0.04
	N	57,545	57,545	64,531
	Std. deviation	1.00	1.00	1.05
Total	Mean	0.00	0.00	−0.03
	N	60,461	60,461	67,492
	Std. deviation	1.00	1.00	1.05

Northern Italy (NUTS 380012 and 380013, Piemont, Valle d'Aosta, Liguria, Lombardia, Alto Adige, Veneto, Friuli-Venezia Giulia, Emilia Romagna)

Southern Italy and Islands (NUTS 380015 and 380016, Abruzzo, Molise, Campania, Puglia, Basilicata, Calabria, Sicilia, Sardegna)

interpretation of the scores with respect to the regions. Table 9.6 shows the result (F=0.134).

The association is not extremely high, but as expected the percentage of respondents counting obedience (and not responsibility) among the five most important is highest in Southern Italy and the Islands and lowest in Baden-Württemberg. The percentages of those who count responsibility (and not obedience) among the five most important items have exactly the reverse order.

Unfortunately differences between Italian provinces cannot be analysed with the European Values Study as the Italian group has only provided region but not province codes for the interviewees. But it is at least possible to compare the two regions of Calabria and Sicily to each other and to the rest of Italy, which leads to the results reported in Table 9.7.

Table 9.7 shows that Sicily is more similar to the rest of Southern Italy and Islands than Calabria, the main difference between the two being the importance of both items which is much higher in Calabria than in Sicily and the rest of Italy and Europe, and the extremely small percentage of interviewees who declare both less important in Calabria as compared to Sicily and the rest of Italy and Europe (all differences between Calabria and the rest are significant on a 5 % level, the differences between Sicily and the rest are not significant with an a of 0.27, whereas the differences between Calabria and Sicily are significant at a level a=0.007, measured with c^2).

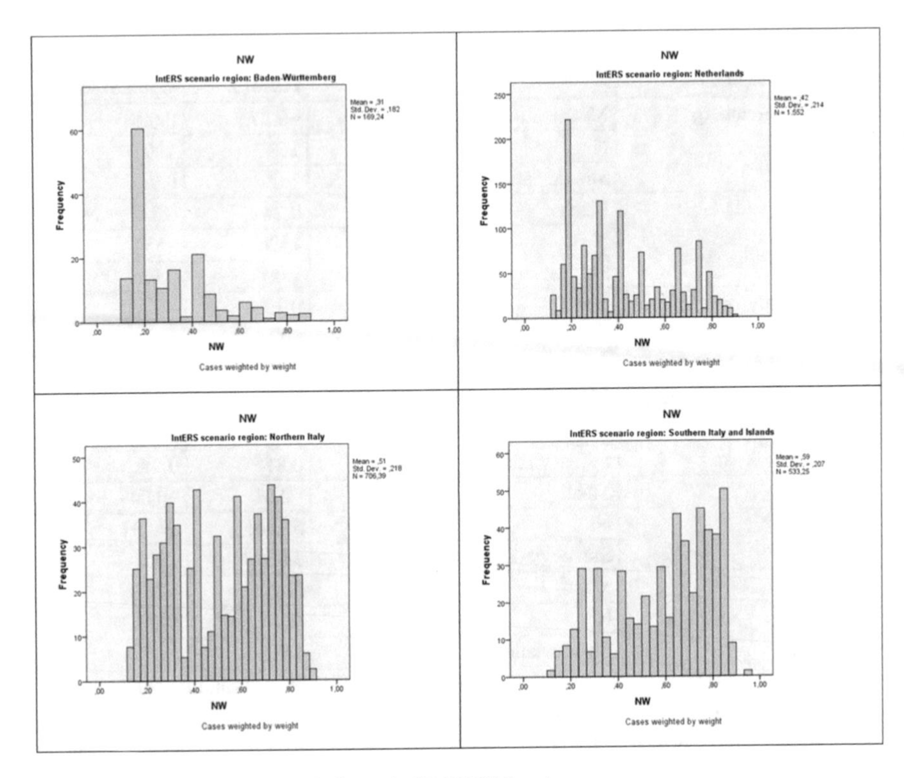

Fig. 9.1 Distribution of the main factor in GLODERS regions

9.2 Prevalence of Crime and Propensity to Denounce Criminals

Another source is Eurobarometer 79.1 (European Commission & Brussels, 2013), carried out between February 23 and March 10, 2013. Although this survey did not address extortion racket systems, some of its results about the attitudes of the interviewees towards corruption can perhaps be used as indicators of the attitudes of people towards paying (or not paying) pizzo (i.e. extortion money) and towards denunciation. The situation of people from whom a bribe is demanded is at least similar to the situation of people from whom a pizzo is demanded, and the decision to denounce will be made according to the same rules in both cases. This is why the relation between the prevalence of corruption and the willingness to denounce corrupt practices can be taken as a surrogate of the respective relation between the prevalence of extortion and the willingness to denounce an extorter.

An analysis of a group of questions referring to corruption-related attitudes yielded three common factors which can be described with the matrix of factor loadings (the coding of the items is 1 = yes and 2 = no) reported in Table 9.8.

Table 9.6 Most important items in GLODERS regions

Crosstabulation of GLODERS regions and Importance of obedience and responsibility

		v173_v180 Importance of obedience and responsibility				Total
		Both items important	Only responsibility important	Only obedience important	Both items unimportant	
Baden-Württemberg	Count	18	133	4	13	168
	%	10.7 %	79.2 %	2.4 %	7.7 %	100.0 %
The Netherlands	Count	355	968	101	113	1537
	%	23.1 %	63.0 %	6.6 %	7.4 %	100.0 %
Northern Italy (as defined in Table 9.5)	Count	159	443	43	59	704
	%	22.6 %	62.9 %	6.1 %	8.4 %	100.0 %
Southern Italy and Islands (as defined in Table 9.5)	Count	159	271	42	59	531
	%	29.9 %	51.0 %	7.9 %	11.1 %	100.0 %
Other	Count	10,769	32,135	6254	10,300	59,458
	%	18.1 %	54.0 %	10.5 %	17.3 %	100.0 %
Total	Count	11,460	33,950	6444	10,544	62,398
	%	18.4 %	54.4 %	10.3 %	16.9 %	100.0 %

Table 9.7 Most important items in Northern and Southern Italy

Crosstabulation of Italian regions and Importance of obedience and responsibility

		v173_v180 Importance of obedience and responsibility				Total
		Both items important	Only responsibility important	Only obedience important	Both items unimportant	
Northern Italy (as defined in Table 9.5)	Count	159	443	43	59	704
	%	22.6 %	62.9 %	6.1 %	8.4 %	100.0 %
Southern Italy and Islands (as defined in Table 9.5)	Count	159	271	42	59	531
	%	29.9 %	51.0 %	7.9 %	11.1 %	100.0 %
Of which: Calabria	Count	27	36	5	1	69
	%	39.1 %	52.2 %	7.2 %	1.4 %	100.0 %
Sicily	Count	27	68	12	17	124
	%	21.8 %	54.8 %	9.7 %	13.7 %	100.0 %
Rest of Southern Italy and Islands	Count	105	168	25	42	340
	%	30.9 %	49.4 %	7.4 %	12.4 %	100.0 %

Table 9.8 Factor analysis of corruption types (varimax rotated component matrix)

	Corruption is not part of business culture	Anti-corruption measures are not effective	EU and national/regional/local institutions are not corrupt
qb15_13 Corruption: polit connections for business	.751		
qb15_14 Corruption: hampering business competition	.747		
qb15_11 Corruption: bribery often the easiest way	.740		
qb15_4 Corruption: part of business culture	.712		.278
qb15_2 Corruption: in national public institutions	.624		.556
qb15_1 Corruption: in local/reg public institutions	.623		.530
qb15_10 Corruption: business and politics too close	.590	−.229	
qb15_5 Corruption: pers affected in daily life	.582		
qb15_7 Corruption: high-level cases not pursued	.467	−.259	
qb15_8 Corruption: government efforts are effective	−.240	.750	
qb15_6 Corruption: enough successful prosecutions		.735	
qb15_12 Corruption: polit party financing supervision		.726	
qb15_15 Corruption: measures are implied impartially	−.220	.675	
qb15_9 Corruption: EU institutions help reducing		.615	−.438
qb15_3 Corruption: in EU institutions			.850

If one compares different regions of Italy and the rest of the European Union along the lines of these factors (whose means are normalised to 0 and whose standard deviations are normalised to 1), this comparison yields the following result from which one can conclude that people in EU 15 minus Italy feel that "corruption is not part of business culture" on a level of one-third of a standard deviation above the mean of EU 28, whereas Northern Italy is one-third of a standard deviation below this mean, and people living in Southern Italy and the Islands have a mean which is even two-thirds of a standard deviation below the EU 15 minus Italy mean. Italy as a whole is approximately 0.46 standard deviations below the EU 28 mean (Table 9.9).

The belief in the effectivity of anti-corruption measures and in the incorruptibility of political and administrative institutions is least in the middle of Italy (Toscana, Umbria, Marche Lazio), not in Southern Italy and the Islands (as one could have thought, perhaps this is an indicator or the resignation there).

The denunciation propensity can be estimated with the help of two questions from the same survey where interviewees were asked:

QB14: I am going to read out some possible reasons why people may decide not to report a case of corruption. Please tell me those which you think are the most important?

We list the proportions of the positive answers in Table 9.10. The percentages in bold show that the reporting culture in Italy differs from the respective cultures both in the EU 15 and in the member states which joined the EU after 1995. What is particularly interesting is that Italians would know where to report, even had no difficulties to prove, but find that "no one does"; they do not believe that reporting is not worth the effort, but they are afraid that there will be no protection. The reason "No one wants to betray anyone" (the omertà reason) is mentioned much less frequently than in the rest of the EU.

The number of interviewees who say that they have reported a case of corruption is too small to yield any significant general results, but at least the difference between Northern and Southern Italy is significant with a Pearson $c^2=7.477$, $a=0.006$ (Table 9.11).

Any regionally deeper analysis (as in the case of the European Values Study) is impossible as the number of interviewees in both Calabria and Sicily was too small in this Eurobarometer.

Nevertheless, an analysis of all regions in Europe can give additional information on the relation between prevalence of corruption and willingness to denounce. For this purpose, the 289 regions represented in the Eurobarometer are weighted with the number of individuals interviewed. Then the correlation between the number of people believing that in their country corruption is fairly widespread and the proportion of interviewees having reported of all who experienced corruption is –0.238 (see also Fig. 9.2). In Fig. 9.2 this is particularly visible for the regions where more than 75 % of the interviewees believe that corruption is widespread in their country: Within these regions the percentage of those who experienced corruption and

Table 9.9 Factor scores in different parts of Europe

Region	N	Corruption is not part of business culture		Anti-corruption measures are not effective		EU and national/regional/local institutions are not corrupt	
		Mean	SD	Mean	SD	Mean	SD
Northern Italy (as defined in Table 9.5)	322	–0.35	0.82	–0.14	1.26	–0.14	0.92
Middle Italy (NUTS 380014, Toscana, Umbria, Marche, Lazio)	159	–0.42	0.77	–0.25	1.31	–0.51	0.86
Southern Italy and Islands (as defined in Table 9.5)	231	–0.65	–0.83	–0.05	1.42	–0.31	0.86
EU 15 without Italy	8177	0.33	1.02	–0.02	0.95	–0.10	1.01
EU members after 1995	5878	–0.40	0.80	0.04	1.03	0.17	0.97
Total	14,766	0.00	1.00	0.00	1.00	0.00	1.00

reported is smaller in regions with higher corruption prevalence—another indicator that victims of corruption do not believe that reporting corruption will have any effect. The interesting exception of this rule is the Italian province of Lombardia where 94.4 % (weighted) of the 159 (unweighted) interviewees believe that corruption is very or fairly widespread in Italy, where 2.2 % (unweighted: three interviewed persons) experienced that a bribe was demanded of which 80 % (unweighted two persons) reported this fact—hence this exception does not actually count as an exception.

If one weights the region with the number of interviewees who have corruption experience, the correlation between the number of people believing that in their country corruption is fairly widespread and the proportion of interviewees having reported of all who experienced corruption is slightly stronger (–0.311).

From the sparse data available on the problem of the relation between falling victim to a crime such as bribe request and denouncing this attempt one can conclude that the prevalence of corruption and the readiness to denounce are negatively correlated, at least where the prevalence is high. For medium and low prevalence the Eurobarometer is not very helpful as the number of relevant cases is too small (2500 in total, only 36 regions with 20 and more interviewees who answered that they had experienced some kind of corruption, and 36 regions with none such interviewee at all).

Table 9.10 Reasons for not reporting corruption in different parts of Europe

	Italy Region Italy North/Centre/South vs. Rest of EU 15 and new members					Total
	Northern Italy (as defined in Table 9.7)	Middle Italy (as defined in Table 9.7)	Southern Italy and Islands (as defined in Table 9.7)	EU 15 without Italy	EU members after 1995	
qb14.1 Do not know where to report it to	54	29	40	3494	2116	5734
	11.5%	**12.7**%	**12.4**%	23.8%	17.5%	
qb14.2 Difficult to prove anything	171	78	111	7514	5479	13,352
	36.5%	**33.6**%	**34.5**%	51.3%	45.3%	
qb14.3 Reporting it would be pointless because those responsible will not be punished	172	87	130	4840	4670	9898
	36.6%	**37.8**%	**40.4**%	33.0%	**38.6**%	
qb14.4 Those who report cases get into trouble with the police or other authorities	88	60	65	2616	3207	6036
	18.8%	25.8%	20.2%	17.8%	26.5%	
qb14.5 Everyone knows about these cases and no one reports them	133	67	100	2595	2957	5852
	28.4%	**29.1**%	**31.1**%	17.7%	24.4%	
qb14.6 It is not worth the effort of reporting it	36	25	20	2829	2405	5316
	7.8%	**10.8**%	**6.3**%	19.3%	19.9%	
qb14.7 There is no protection for those who report corruption	178	86	140	4141	4093	8638
	38.0%	**37.1**%	**43.7**%	28.2%	33.8%	
qb14.8 No one wants to betray anyone	38	10	11	2838	1860	4757
	8.1%	**4.4**%	**3.5**%	19.4%	15.4%	
qb14.9 Other (SPONTANEOUS)	5	2	1	291	133	431
	1.0%	1.0%	0.2%	2.0%	1.1%	
qb14.10 None (SPONTANEOUS)	23	8	7	417	93	547
	4.9%	3.3%	2.1%	2.8%	0.8%	
qb14.11 Do not know	24	5	18	405	325	777
	5.1%	2.4%	5.5%	2.8%	2.7%	
Total	469	231	321	14,661	12,105	27,786

Percentages and totals are based on respondents

Table 9.11 Frequency of corruption reporting in Northern and Southern Italy

	Corruption experience reported		Total
	1 Yes	2 No	
Northern Italy	7	12	19
Southern Italy and Islands	1	23	24
Total	8	35	43

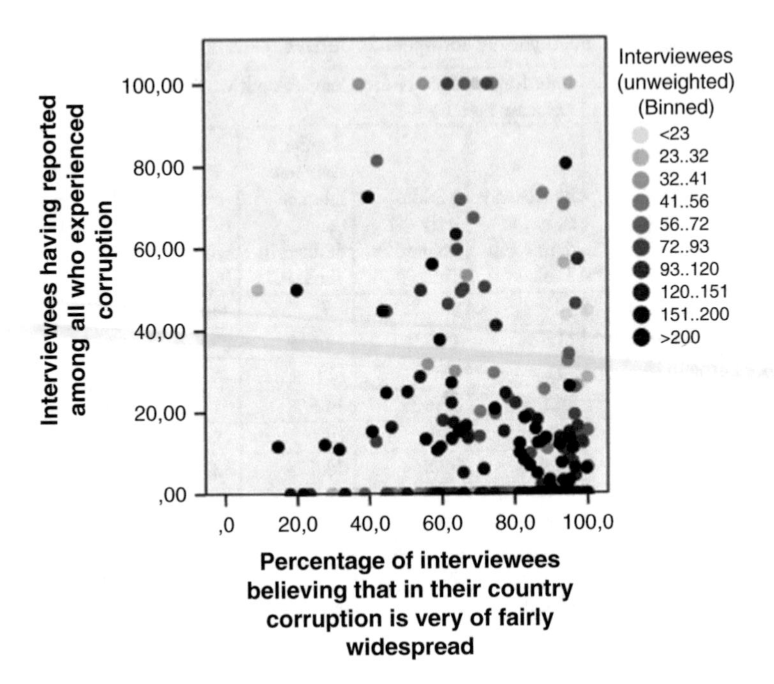

Fig. 9.2 Scattergram of prevalence of corruption vs. willingness to report corruption

9.3 Behaviour Related to Menace, Assault, and Aggression in Italy

The Italian National Institute of Statistics (ISTAT) has carried out several surveys about citizens' security, the last wave having been in the field in 2008 (Istituto Nazionale di Statistica, 2008–2009). At least two sections (10A and 10B) seem appropriate to describe the setting in which victims of extortion decide whether they denounce or pay the requested amount—although the questions in the two mentioned sections are about unspecified menaces and aggressions.

In this survey 60,001 persons were interviewed by telephone. 2187 interviewees reported that they had experienced a case of menace or of assault or aggression during the past 3 years. 216 or 9.9 % of these cases were reported to the police. Two questions allow for a differentiation of both victims and offenders. Those interviews who reported to be entrepreneurs/imprenditori, professionals/liberi professionisti, self-employed/lavoratori in proprio, etc. could be called "typical targets", while the others (employees of different categories) could be called implausible victims. Among the offenders there are those whom the interviewees described as strangers ("estraneo") of loose acquaintances ("Una persona che conosceva di vista") or whom they did not want to describe ("non risponde")—these offenders could be called possible offenders, whereas the other offenders (i.e. members of the family, colleagues, or near neighbours) are rather implausible offenders.

Table 9.12 Offences similar to extortion by regions, target type and offender type, and percentage of denounced offences

Regions		Implausible targets, implausible offenders	Implausible targets, possible offenders	Typical targets, implausible offenders	Typical targets, possible offenders	Total/per cent
Campania, Puglia, Calabria, Sicilia	N	260	121	188	90	659
	%	39.5 %	18.4 %	22.5 %	13.7 %	100 %
	[a]	6.2 %	26.4 %	2.1 %	22.2 %	10.4 %
Rest of Italy	N	770	339	295	124	1528
	%	50.4 %	22.2 %	19.3 %	8.1 %	100 %
	[a]	4.0 %	23.2 %	5.4 %	14.5 %	9.9 %

[a]Percentage of denounced offences; the numbers do perhaps not add up to the total as the interviewees are weighted, and the absolute numbers are rounded

If one analyses the data according to these criteria, Table 9.12 shows the results.

It becomes visible that those offences which can be interpreted as extortions are much more frequent in Campania, Puglia, Calabria, and Sicily than in the rest of Italy, and it is also true that in the mentioned regions denunciations are more frequent. Generally speaking, entrepreneurs and professionals seem to denounce slightly less often than the others (8.9 % vs. 10.6 %, but with Fisher's exact test the significance level is only 0.056). Offences by strangers or loose acquaintances are much more often denounced than those committed by family members, colleagues, or close acquaintances (22.1 % vs. 4.4 %, significance level of Fisher's exact test is less than 0.0000005). Denunciation propensity is slightly higher in the named regions than in the rest of Italy (11.1 % vs. 8.4 %, significance level of Fisher's exact test is 0.212).

More detailed analyses are not very reasonable as the prevalence of all these offences is still moderate, and in spite of a very large sample of 60,001 cases the subsample of offences similar to extortion (i.e. menaces and aggressions committed by possible offenders towards typical targets as defined above) is with 214; see Table 9.12: Offences similar to extortion by regions, target type and offender type, and percentage of denounced offences—too small.

9.4 The Sicily and Calabria Database

This subsection will use a database compiled from 629 cases from all provinces of Sicily and Calabria (Frazzica et al., 2015)—see also Chap. 7—which allows for comparisons between provinces and for making more than one empirical case available for a comparison to simulation output (as each simulation run can be considered as a history produced in a fictitious province with its initial conditions and parameter settings).

Table 9.13 Frequency of extortion denouncement in provinces in Southern Italy

	All extortions			Denounced				
		Completed		1 no		2 yes		
Province	Attempted	Number	Per cent	Number	Per cent		Of which with conviction	Total
Agrigento	30	51	63.0	58	71.6	23	20	81
Bergamo	0	1	[a]	1	[a]	0	[a]	1
Campania	1	0	[a]	0	[a]	1	[a]	1
Catania	17	93	84.5	84	76.4	26	25	110
Messina	21	66	75.9	66	75.9	21	21	87
Milano	1	1	[a]	2	[a]	0	[a]	2
Palermo	45	91	66.9	114	82.6	24	MD	138
Ragusa	2	13	86.7	15	[a]	0	[a]	15
Reggio Calabria	27	56	67.5	76	87.4	11	MD	87
Siracusa	10	50	83.3	52	86.7	8	8	60
Trapani	28	14	33.3	32	76.2	10	6	42
Vibo Valencia	0	1	[a]	1	[a]	0	[a]	1
Total	182	437	70.6	501	80.2	124	(80)	625

The per cent columns are used for Figs. 8.2, 12.1, and 12.2
[a]Absolute numbers too small, MD: missing data, four cases have missing data for the denunciation variable and are therefore excluded

A bivariate analysis also shows that about 80 % of all extortions reported in this database were never reported by the victims (see Table 9.13), but in those provinces where our data are more or less complete, nearly all of the denunciations led to a conviction.

Table 9.14 also shows that only a minority of extortion requests were refused (53 out of 224, 23.7 %) and that out of these in only 16 cases (30.2 %) the information of the police came directly from the victim; in 14 more cases (26.4 %) information from the victim led to denouncement, but even in the refusal case 23 extortions (43.4 %) were found out by the police without any help from the victim. When the victim's reaction was acquiescence or connivance, there was no direct denouncement (the eight cases with acquiescence where the first information came from the victim seem to have remained below the threshold of denouncement), of course; from these 161 cases, 138 were detected by police activities (84.7 %), and help from the side of the victims was only given in 15.3 % of these cases. Differences between the three provinces where we have data about both variables are small.

If one only considers the victim reaction and the fact whether there was a denouncement (here we have sufficient data for all the provinces in Southern Italy), the analysis yields the following picture (Table 9.15): The association between the two variables is very high with the exception of the province of Catania (both Cramer's V and the uncertainty coefficient are highly significantly different from 0.0).

Table 9.14 Reaction of the extortion victim in provinces of Southern Italy

Province	Extortion reaction	Information acquisition mode (simplified, 3 values)			Total
		1 victim	2 victim and police or police collaborator	3 police or police collaborator	
Palermo	Acquiescence	3	13	72	88
	Connivance	0	0	8	8
	Resistance	10	7	10	27
	More complicated cases	1	1	6	8
	Total	14	21	96	131
Reggio Calabria	Acquiescence	4	2	46	52
	Connivance	0	0	5	5
	Resistance	5	1	11	17
	More complicated cases	0	0	1	1
	Total	9	3	63	75
Trapani	Acquiescence	1	0	4	5
	Resistance	1	5	1	7
	More complicated cases	0	0	1	1
	Total	2	5	6	13
Total	Acquiescence	8	15	125	148
	Connivance	0	0	13	13
	Resistance	16	14	23	53
	More complicated cases	1	1	8	10
	Total	25	30	169	224

Another high correlation exists between the victim reaction and the fact whether the extortion was actually committed or only attempted (see Table 9.16).

Again the association is very high except in the case of the province of Siracusa (all uncertainty coefficients with the exception of Siracusa's and all Cramer's V are again significantly different from 0.0).

If one goes further back in a presumable chain of causalities the following can be found (Table 9.17): Repeated payments are more often successfully requested (85.4 %) than one-off payments (63.5 %); the request for both at the same time is very often (83.7 %) successful. The effect sizes (Cramer's V=0.230, uncertainty coefficient with the success of the operation dependent=0.047) are only moderate, but both are highly significant, but in most provinces the 0.05 significance level is not reached.

Surprisingly the reported intimidation only has a small effect on the success of the operation of extortion (Cramer's V is 0.117, uncertainty coefficient is 0.015, both are significant, a is 0.004 and 0.001, respectively, but for such a small effect the significance is rather irrelevant).

Table 9.15 Relation between victim reaction and denouncement in provinces of Southern Italy

Province	Extortion reaction	Denounced 1 no	Denounced 2 yes	Total	Per cent resistance	Cramer's V	Uncertainty coefficient
Agrigento	Acquiescence	56	12	68	14.8	0.587	0.273
	Connivance	1	0	1			
	Resistance	1	11	12			
	Total	58	23	81			
Catania	Acquiescence	70	22	92	9.1	0.190	0.0470
	Connivance	8	0	8			
	Resistance	6	4	10			
	Total	84	26	110			
Messina	Acquiescence	65	13	78	8.0	0.538	0.240
	Connivance	1	1	2			
	Resistance	0	7	7			
	Total	66	21	87			
Palermo	Acquiescence	83	4	87	20.9	0.622	0.368
	Connivance	5	3	8			
	Resistance	10	17	27			
	More complicated cases	8	0	8			
	Total	106	24	130			
Reggio Calabria	Acquiescence	49	3	52	22.7	0.446	0.219
	Connivance	5	0	5			
	Resistance	10	7	17			
	More complicated cases	1	0	1			
	Total	65	10	75			
Siracusa	Acquiescence	48	4	52	11.7	0.469	0.199
	Connivance	4	3	7			
	Resistance	0	1	1			
	Total	52	8	60			
Trapani	Acquiescence	18	0	18	50.0	0.559	0.370
	Connivance	2	0	2			
	Resistance	11	10	21			
	More complicated cases	1	0	1			
	Total	32	10	42			
Total	Acquiescence	403	58	461	16.7	0.418	0.148
	Connivance	26	7	33			
	Resistance	43	58	101			
	More complicated cases	10	0	10			
	Total	482	123	605			

Table 9.16 Relation between reaction of victim and success of the extortion attempt in provinces of Southern Italy

Province	Extortion reaction	Attempted and completed extortion			Total	ϕ	Uncertainty coefficient
		1 only attempted	2 only partly completed	3 completed			
Agrigento	Acquiescence	18		50	68	0.501	0.199
	Connivance	1		0	1		
	Resistance	11		1	12		
	Total	30		51	81		
Catania	Acquiescence	7		85	92	0.742	0.477
	Connivance	0		8	8		
	Resistance	10		0	10		
	Total	17		93	110		
Messina	Acquiescence	13		65	78	0.537	0.240
	Connivance	1		1	2		
	Resistance	7		0	7		
	Total	21		66	87		
Palermo	Acquiescence	2	1	84	87	0.643	0.726
	Connivance	3	0	5	8		
	Resistance	27	0	0	27		
	More complicated cases	7	0	1	8		
	Total	39	1	90	130		
Reggio Calabria	Acquiescence	0	0	50	50	0.708	0.915
	Connivance	0	0	5	5		
	Resistance	16	1	0	17		
	More complicated cases	1	0	0	1		
	Total	17	1	55	73		
Siracusa	Acquiescence	9		43	52	0.327	0.114
	Connivance	0		7	7		
	Resistance	1		0	1		
	Total	10		50	60		
Trapani	Acquiescence	7		11	18	0.663	0.400
	Connivance	0		2	2		
	Resistance	20		1	21		
	More complicated cases	1		0	1		
	Total	28		14	42		

(continued)

Table 9.16 (continued)

Province	Extortion reaction	Attempted and completed extortion			Total	ϕ	Uncertainty coefficient
		1 only attempted	2 only partly completed	3 completed			
Total	Acquiescence	56	1	402	459	0.507	0.410
	Connivance	5	0	28	33		
	Resistance	96	1	4	101		
	More complicated cases	9	0	1	10		
	Total	166	2	435	603		

Table 9.17 Extortion types and success of extortion attempt

Periodic or una-tantum request	Attempted or completed extortion		Total
	1 no, only attempted	2 yes, completed	
One-off	114	198	312
Periodically	6	35	41
Both	38	195	233
Total	158	428	586

Table 9.18 Intimidation and success of extortion attempt

Intimidation	Attempted or completed extortion		Total
	1 no, only attempted	2 yes, completed	
1 yes	179	401 = 69.1 %	580
2 no	2	29 = 93.5 %	31
Total	181	430	611

Table 9.19 Kind of demand and success of extortion attempt

	Attempted or completed extortion		Total
	1 no, only attempted	2 yes, completed	
Money	81 = 33.2 %	163 = 66.8 %	244
Other (goods, services, staffing, unknown)	33 = 17.7 %	153 = 82.3 %	186
Total	114	316	430

The intimidation (Table 9.18) seems to have rather the opposite effect, but with only a very small number of extortions without intimidation this is not relevant either.

Whether only money is requested or something else (goods, services, staff, or any of these combined with money) makes a small difference (which is statistically significant at 0.0005 level, but the effect measured with Cramer's V 0.173 is rather modest) (Table 9.19).

For about one-third of the cases no information was available.

Given the fact that the empirical data do not cover the whole theoretical model, as much information backing the micro-specification is more or less unobservable (which is particularly obvious in the case of the extorters), validation of the theory and the simulation model derived from the theory is difficult. This will be discussed in more detail in Sect. 12.1.

References

European Commission, Brussels. (2013). *Eurobarometer 79.1. ZA5687 Data file Version 3.0.0.* Abgerufen am 26. 2 2016 von GESIS Data Archive, Cologne: doi:10.4232/1.12448.

EVS. (2008). *European Values Study 2008: Integrated Dataset (EVS 2008). ZA4800 Data file version 3.0.0.* Abgerufen am 26. 2 2016 von GESIS Data Archive: doi:10.4232/1.11004.

Frazzica, G., Punzo, V., La Spina, A., Militello, V., Scaglione, A., & Troitzsch, K. G. (2015). *Sicily and calabria extortion database.* (GESIS, Hrsg.) Von Datorium: http://dx.doi.org/10.7802/1116 abgerufen.

Istituto Nazionale di Statistica. (2008–2009). Sicurezza dei cittadini.

Part IV
The Criminal Organisation

Chapter 10
Text Data and Computational Qualitative Analysis

Martin Neumann and Ulf Lotzmann

10.1 Introduction

Parts II and III of this book focussed on the interaction between state and civil society in an extortion racket system. In this part of the book we turn to the third corner of the typology discussed in Chap. 3, the criminal organisation itself. We approach the investigation of the modes of organisational behaviour in the crime field from its reverse angle: the breakdown of a criminal network in an escalation of intra-group violence. This chapter discusses background literature, the empirical basis of the investigation into the criminal organisation and a conceptual model extracted from the data. Chapter 11 presents an agent-based model resulting from this empirical basis.

10.1.1 Criminal Collapse: Analysis of the Breakdown of a Criminal Network

Specific problems of criminal organisations are only rarely investigated (comp. Diesner, Frantz, & Carley, 2005 for an example). However, as criminal organisations operate outside the legal world in which social order is secured by the state monopoly of violence (Sofsky, 1996) they face specific problems: In fact, while scientific research approaches criminals mainly as offenders, they are also

M. Neumann, Ph.D. (✉) • U. Lotzmann
Department of Computer Science, Institute for Information Systems Research, University of Koblenz-Landau, Universitätsstr 1, Koblenz 56070, Germany
e-mail: maneumann@uni-koblenz.de; martneum@freenet.de; ulf.lotzmann@uni-koblenz.de; ulf@uni-koblenz.de

C. Elsenbroich et al. (eds.), *Social Dimensions of Organised Crime*, Computational Social Sciences, DOI 10.1007/978-3-319-45169-5_10

potential victims (Putten, 2012). In case of a criminal offence reliance on the law enforcing agencies of the state such as the police comes along with high costs for criminals. For instance they might be subject to a criminal prosecution themselves or need to be protected against their former criminal comrades. This provides a source that criminals themselves are highly vulnerable against criminal offences. Therefore examining malfunction of a criminal group sheds light on the conditions for their organisational behaviour. For this reason we apply an organisational science perspective to the investigation of the intra-organisational norms in the criminal world (Gottschalk, 2010; Weick, 2007). For instance, in our examination of a case of extortion it turned out that the very term 'extortion' implies a perspective of criminal law on a certain kind of behaviour. However, from an organisational science perspective the very same activity appears as a run on the bank (Merton, 1968). Certainly this does not entail any justification of criminal activities but rather has to be perceived as an analytical concept for disentangling criminal norms.

The state of the art of research on the organisation of organised crime can be disentangled along the question of the degree of organisational growth and rationalisation of labour (von Lampe, 2015). Starting point for the academic debate has been the mafia as the paradigm of a professional, hierarchically organised crime syndicate (La Spina, 2005; Paoli, 2003). Whereas Cressey (1969, 1972) developed the thesis that like legal companies also criminal organisations tend to grow and develop an increasing rational management of labour, Reuter (1983) argued for the contrary thesis that particular conditions of criminal markets favour small and local enterprises. Chang, Lu, and Chen (2005) developed a model to determine organisational size as a variable dependent on environmental conditions. Namely, criminal organisations face a trade-off between efficiency and security (Morselli, Giguere, & Petit, 2006). While organisational growth, structural differentiation, and a rational organisation of the group management might increase returns, organisational growth comes at the cost of increasing danger of being detected (von Lampe, 2015). Small and local groups provide more security against criminal prosecution. Thus the specific condition of covertness shapes the kind of interactions and relations within and beyond the criminal organisations. It is argued that covertness favours flexible and adaptive networks without hierarchical relations. These might quickly emerge and dissolve for temporarily taking advantage of criminal opportunities (Klerks, 2002; Krebs, 2002; Morselli, 2009; Sparrow, 1991). For instance in the case of New York's heroin market Natarajan (2006) found only small groups of entrepreneurs rather than big criminal syndicates. Thus emphasis has shifted from studying organisations to processes organising (Hobbs, 2001). However, current research is focused on a static picture. Predominantly the structure of criminal organisations is perceived as a kind of rational or evolutional adaptation, may it be to environmental conditions (favouring small networks) or conditions for the efficiency of production (favouring big syndicates). In contrast here we examine the dynamics of relations within a criminal group. Investigating the dynamics sheds light on the pitfalls in which criminal groups might be trapped. These are not based on deliberate decision but can be perceived as unintended consequences of actions. In contrast to the picture of adaptive flexibility of small networks drawn in the literature, the case examined here reveals the negative side effects that organisational growth has on small and flat organisational structures.

10.2 Data Basis

Empirically, the chapter investigates the collapse of a gang of criminals involved in drug trafficking and laundering illegal money. The basis of the data are transcripts of police interrogations of witnesses as well as suspects in a number of cases that were related with each other insofar as a core group of persons were involved in all these cases. This core group consisted of ca. 20–30 persons. Some of these persons knew each other already for several decades, partly also by a record of co-offences in a longtime criminal career. In contrast to the Sicilian Cosa Nostra, the group had no hierarchical structure or formal positions such as a *capo di famiglia*, i.e. the head of a certain sub-unit of the Cosa Nostra operating in a certain district. Whereas the Cosa Nostra is a professional organisation (La Spina, 2005; Neumann & Sartor, 2016; Scaglione, 2011) the group subject to these police investigations was more of a network of old friends. While certainly some individuals gained more prestige than others the structure of the group did not consist of positions with managerial authority or right of command. However, at least for a decade the group operated extremely successfully in the drug market. The groups were formed presumably in the early 1990s and made a lot of money in particular with ecstasy. They made a profit equivalent to several hundred millions of Euros that had been laundered in highly professional, worldwide financial transactions (Neumann & Sartor, 2016).

However, in the early to mid-2000s the group collapsed in an escalation of violence. An informal network cannot terminate as, e.g. a legal company declaring bankruptcy. So collapse means in this case that the business relations terminated, either because they killed each other or because of a loss of trust, partly due to murders and other acts of violence including kidnapping, intimidation, and extortion. Collapse of trust is essential for the breakdown of the group. As already in legal organisations trust is essential for the efficiency of labour relations (Colquitt, LePine, Piccolo, & Zapata, 2012; Colquitt & Rodell, 2011), this holds even more in the case of illegal organisations operating outside the state monopoly of violence. In the legal domain organisations can at least ultimately rely on the state monopoly of violence: As labour relations are contractual relations norm enforcement can be delegated to the third party of the court. In criminal organisations recourse to the court is impossible. Thus a criminal organisation needs to rely on the commitment of the members to the organisation. For this reason, trust is essential. For instance, in case of money laundering black collar criminals need to hand over the money to their partners and trust that they will get the return of investment back from the trustee. In a covert organisation this cannot be secured by formal contracts. Once trust is corrupted the business relation breaks apart. In fact the collapse of the group triggered massive violence, including a large number of assassinations. For instance, three murders happened within one week. In turn the violence fostered further breakdown of trust. The escalation of violence has been described by involved persons as a 'corrupt chaos' governed by a 'rule of terror' in which 'old friends are killing each other'. The notion of 'chaos' indicates that seemingly the 'terror' was not governed by an individual such as Nero burning Rome but—from the perspective from inside—by an invisible hand. The involved persons could no longer keep track of the complexity of incidences. This is an emergent phenomenon in which the macro level of the situational complexity generates a perception of the situation as a 'corrupt chaos' on the micro level of the involved individuals. This motivates the research question of the data analysis: dissecting the mechanisms of the chaos on a level of fine-grained individual interactions. For this purpose a phenomenological description is applied.

10.3 Methodological Approach

The analysis is based on several police investigations in which numerous interrogations are documented. Police interrogations can be described as situations of dialogical conversation. An in-depth analysis of subjective meaning attributed to certain situations brings the empirical analysis very close to the subjective perception of the actors. Certainly interrogations are artificial situations which might be alien for the respondents who might answer strategically or simply lie. Moreover the talk is guided by certain interests of the police. In this case for instance, the police investigations focused on persons related to money laundering and less on drug production. This uncertainty is typical for criminological data (Bley, 2014). On the other hand police interrogations differ from court files in which respondents can be put under oath. Moreover police interrogations are confidential. Many of the respondents were witnesses as for instance relatives of victims who were themselves interested in elucidation of the cases. This gives their statements certain credibility. For this reason the data provides a rather good basis for analysing the cognition of a certain situation as 'corrupt chaos'. The aim is to infer hypothetical, *unobservable* cognitive elements from *observable* actions and statements to analyse cognitive mechanisms that motivates action in very confused and opaque situations.

Investigating subjective perceptions calls for an interpretive research methodology. For this purpose the research draws methodologically upon a Grounded Theory approach. In a first step the data was loaded into MaxQDA (www.maxqda.de) as a tool for qualitative text analysis and text passages were annotated which then were summarised into codes deriving concepts from data. Concepts stand for classes of objects, events or actions which have some major properties in common. This is a classical open coding of a Grounded Theory (see Corbin & Strauss, 2008). However, in a second step the research diverged from classical Grounded Theory. The coding derived with MaxQDA served as the basis for concept relation identification with the CCD tool which is software for creating a conceptual model of the process (Scherer, Wimmer, & Markisic, 2013; Scherer, Wimmer, Lotzmann, Moss, and Pinotti, 2015). This departs from a classical Grounded Theory approach by making use of an abstract framework of condition-action sequences (Lotzmann & Wimmer, 2013). The web of

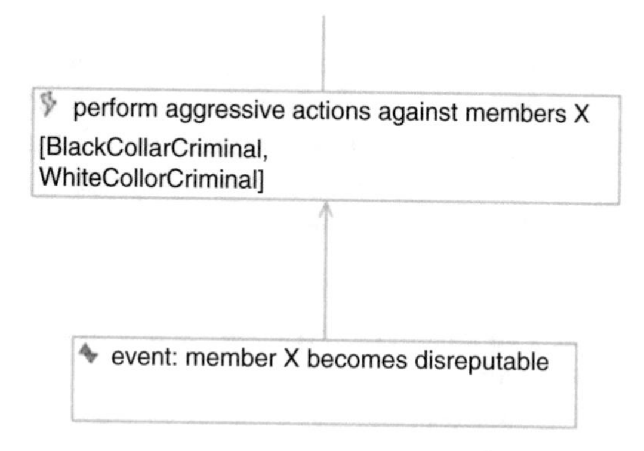

Fig. 10.1 An example of a condition-action sequence

interrelated sequences is denoted as an action diagram. The concept of condition-action sequences is an a-priori methodological device to identify social mechanisms on a micro level of individual (inter-)action. Any process is initiated by a certain condition which triggers a certain action. This action in turn generates a new state of the world which is again a condition for further action. Broadly speaking a mechanism is a relation that transforms an input X into an output Y. A further condition is a certain degree of abstraction which becomes evident by a certain degree of regularity, i.e. that under similar circumstances a similar input X* reveals similar outputs Y*. In the social world it is typically an action which relates X and Y (Hedström & Ylkoski, 2010). This is assured by the concept of event action sequences. Whereas the data describes individual instantiations the condition-action sequences represent general event classes. For instance, in our case one condition is denoted as 'return of investment available'. This triggers the action to 'distribute return of investment'. Obviously this describes classes of events. Return of investment might be rental income as well as, e.g. purchasing of companies. However, empirical validity is ensured by tracing the individual condition-action sequences back to the coding derived with MaxQDA (Neumann & Lotzmann, 2014; Lotzmann, Neumann, & Moehring, 2015). This methodology enables controlled generalisation from the case which provides a proof of existence of the inferred mechanisms. In more technical terms a condition-action sequence looks like displayed in Fig. 10.1.

The diagram represents a condition-action sequence. The box with a red flag represents an event. The action is represented by a box with a yellow flag. Moreover, in brackets we see the possible types of agents that can undertake the action. The arrow represents the relation between the event and the action. This is not a deterministic relation. However, the existence of the condition is necessary for triggering the action. Once an action is performed a new situational condition is created which again triggers new actions. This is indicated by the stripe leaving the box which represents the action. The sequence does not represent a concrete event in space and time but a class of possible relations. Moreover, the CCD tool creates a code template which can be implemented in a simulation model. The model will be presented in the next chapter (Chap. 11). On the other hand traceability to the empirical data is secured by annotations that refer to the open coding performed in the first step of the analysis. In sum, a web of condition-action sequences is generated that represents the conceptual model of the data. Developing the conceptual model is an iterative process: first the individual condition-action sequences need to be consistent with empirical domain knowledge. Second, the overall web of relations needs to provide a meaningful big picture that is sufficient to represent the overall corpus of the data. Therefore the development of the conceptual model has been a participatory modelling process, i.e. stakeholder knowledge of police experts went into the model. Several developmental stages of the model have been discussed with stakeholders until they perceived it as valid. This is an equivalent to the concept of theoretical saturation in a Grounded Theory approach (see Corbin & Strauss, 2008). Finally note that the data basis of interrogations allows including cognitive conditions (such as 'fear for life') and actions (such as 'member X interprets aggressive action'). This is an important feature to achieving at a thick description from a situational phenomenology. For understanding the chaotic terror it is essential to retrieve the meaning attributed to particular situations, observable at a phenomenological level.

10.4 Conceptual Model

In this section the conceptual model of the data and its empirical trace will be presented. The description concentrates on the relation between 'black collar criminals', involved in drug trafficking, and 'white collar criminals' responsible for money laundering. This means that the production and distribution of drugs, i.e. the source of the illegal money, is not taken into account. The conceptual model is realised in the action diagram of the web of condition-action sequences. In the next section the mechanisms of the collapse at micro level of single actions are investigated. In the next section a theoretical analysis of the conceptual model will discuss the mechanisms on the macro level of the structural properties of the criminal group which can be revealed from the micro-level analysis. First it has to be noted that in the investigated relations three kinds of actors are involved:

- 'Black collar criminals' who gained illegal money in the drug business.

'White collar criminals' with a good reputation in the legal society in order to be able to invest huge amounts of money in the legal market. These might not have a long record of criminal offences but might be pushed towards criminal behaviour on course of interactions (Gross, 1978).

So-called straw men which played a decisive role in concealing the source and target of the money flow.

Once the data has been transformed in an action diagram five phases in the process of the collapse can be distinguished in the analysis of the action diagram:

Ordinary business of money laundering: This is the status quo before the collapse took place. Note that production and distribution of drugs is not investigated.

A *crystallising kernel of mistrust* disturbing the ordinary business, initiating the collapse.

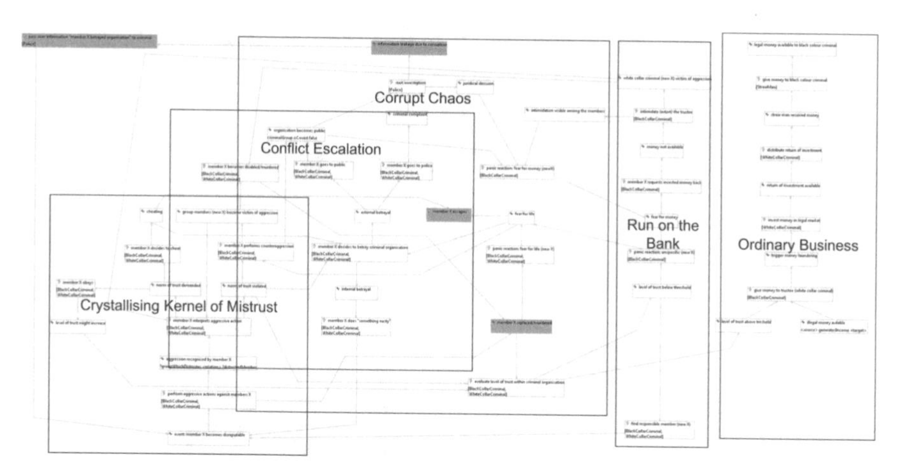

Fig. 10.2 Overview of the action diagram of the process

If the mistrust cannot be encapsulated a spreading of mistrust through the group generates a *conflict escalation* which finally leads to what has been denoted in terms of witnesses as a *Corrupt chaos*, including a *run on the bank*. This was part of the 'corrupt chaos'. However, it can be analytically distinguished because the financially oriented relations of conditions and subsequent actions remain separated from the purely existential violence.

In Fig. 10.2 an overview of the condition-action sequences that could have been identified in the data is provided. Note that these sequences generalise from the empirical instantiations. Thus they are social mechanisms that are inferred from the data. However, transparency is ensured by tracing these mechanisms back to the empirical observations from which they are derived.

What can be seen from the overview is the fact that the relations in the ordinary business and the run on the bank, i.e. those actions that are related to financial transactions, remain rather isolated, with rather few links to the web of the other actions. In contrast, the actions in the process of conflict escalation are strongly interrelated. Next, the individual elements (from the ordinary business to the run on the bank) will be described in detail. First, the individual condition-action sequences of the process of the ordinary business of money laundering are considered. We describe the individual sequences and provide text passages in the police interrogations from which these sequences are derived.

10.4.1 Ordinary Business of Money Laundering

The process of money laundering, described in Fig. 10.3, starts with two conditions: Obviously, illegal money must be available. However, black collar criminals invested a huge amount of money in the business of white collar criminals. In the absence of formal contracts which are secured by the possibility that claims can be enforced by

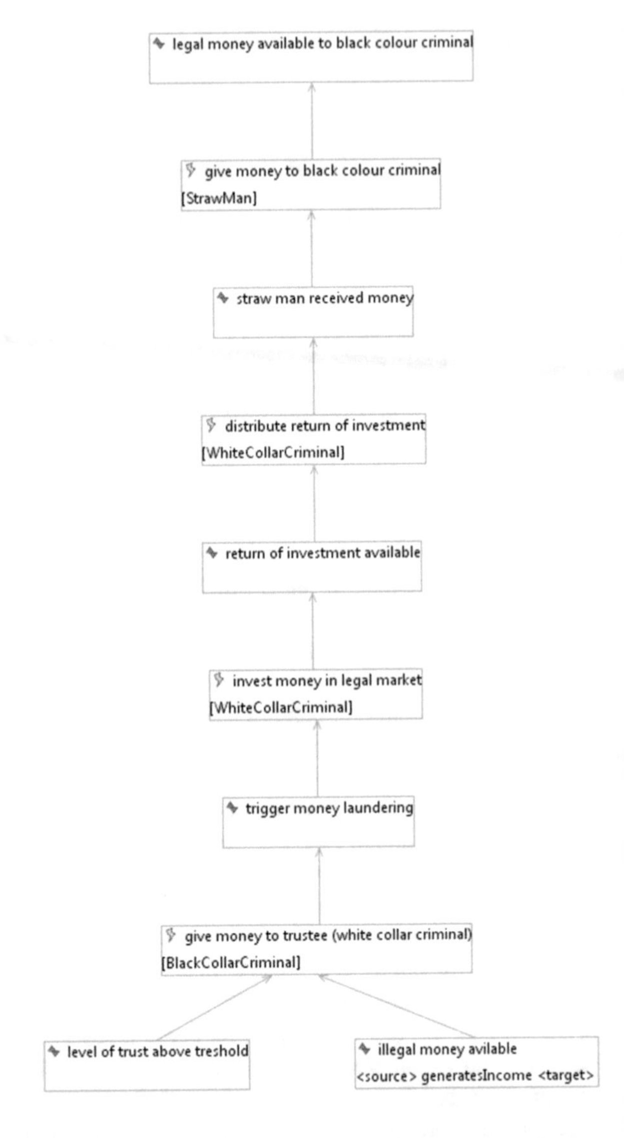

Fig. 10.3 Money laundering

legal action, also trust is required in order to trigger a process of money laundering. These two conditions are inferred from statements in the police interrogation, for which the following two citations are exemplary.[1] The level of trust is expressed in the following statement of a witness, in which O1 is a black collar and V01 a white collar criminal,[2] that money was available is documented in the second report:

[1] These are open codings derived with MaxQDA which are then inserted as annotations in the CCD framework.

[2] For reasons of protection of private data, names are anonymous and no reference to the source in the police interrogations is provided.

O1 and V01 seem to be friends for me.

In the period between 1990 and Feb 14, 1992 police investigations had been undertaken. These revealed a criminal organisation concerned with drug trafficking. The report from June 1992 estimated the income and the costs. It is estimated a transaction volume of nearly 300 million.

If these two conditions are fulfilled a process of money laundering is triggered. In this case illegal money is given to a trustee with a legal business who invests the money in the legal market. The trustee is the link between the illegal and the legal world. That illegal money has been available is testified by the following statement:

… inserted a significant value of black money in the structure of the company of V01.

The money that had been inserted in the company of V01 has been invested in the legal market, as testified in the following statement:

At the moment I paid 800,000 in the firm which are now worth several millions through legal trade.

The investment of the money triggers the redistribution of the now legal money back to the black collar criminals. However, it turned out that for the concealment of source and target of the money third parties had extensively been used. These need to be individuals which are not in the first instance visible parts of the criminal group but nevertheless are trusted by group members. We call them straw men.

Finally, V01 paid 59 million. The cash money had been invested through a construction in Curacao. Here the brother of V01 played a decisive role.

The brother is but one example. Another one had been, e.g. the girlfriend of a criminal. However, this example makes clear the two functions of (a) not being visible as part of the group but nevertheless, (b) being highly trusted by group members. Here the family ties play a decisive role.

10.4.2 A Crystallising Kernel of Mistrust

This process could have gone on without any specific terminal point. However, factually at some point in time a crystallising kernel of mistrust invaded the group. Obviously, this is a contingency of the investigated data: The interrogations are based on the fact that the group became visible and factually the group became visible only in and through the process of its collapse. This is a kind of happenstance. In particular the individual events remain contingent. The story of this particular case will be developed in the textual annotations below. These gave rise to the identification of the mechanisms of the decline of the group. However, it has to be noted that it is rather likely that in the course of time some such events happen that trigger follow-up actions. For this reason the conditions have been specified in the condition-action sequences in a very general way: It is simply stated that some member of the group becomes disreputable. In the first instance, this is due to the limits of the data. In the interrogation it cannot be identified unequivocally why and how this happened. In the data only the follow-up steps can be found; that is, it is a theoretical inference that someone became disreputable. However, first, this is a very general condition which makes it rather likely that some point of time it will occur. Second, in a group some form of conflict resolution is needed. The crucial question is how the group handles conflicts. This is a critical juncture for the stability of the group.

As justified above, the starting point of the process outlined in Fig. 10.4 is treated as an external event, namely that a member of the group becomes disreputable. This may be due to several reasons: for example a member may become too greedy. Once this event happened it calls for a mechanism of conflict resolution. Conflict resolution might trigger an act of aggression against this member. This might be an attempt to sanction this member or motivated by some causes such as, for instance, simply anger or irritation about him or her. However, it may also be the case that mistrust is based on other reasons or that the motivation is based on self-interest or simply remains unclear. In the following some examples of how an aggression might look like in the context of a criminal group are provided:

> An attack to the life of M.
>
> O1 had V01 in his grip. He shall do as told otherwise his family would have a problem.
>
> …O5 came to my house in order to say that at 8 in the evening I should come to the forest. This is standard: intimidate and request for money.

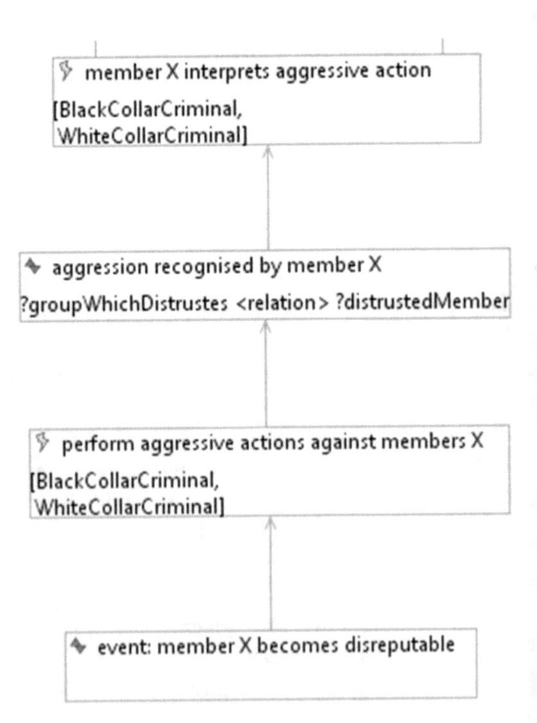

Fig. 10.4 Crystallising kernel of mistrust

The aggression in these examples is of very different severity. Obviously murder is a severe aggression. It shall be noted that assassination might be motivated by several reasons, ranging from greediness to death penalty. In fact, M. survived the attack but had been killed some years later because he had been accused of stealing drugs. The latter can be interpreted as execution of a death penalty. In the other two examples the objective of the action is not the liquidation of the victim. If effective, the aggression is recognised by the victim. This triggers reasoning on the aggression. In contrast to (successful) murder, the aggression in the two other cases is intended to initiate certain behaviour or behaviour change respectively. In the second example O1 'shall do as told', whereas in the third one the objective of the intimidation is a 'request for money'. Recognising the aggression triggers a crucial cognitive process: namely, interpreting the possible motivation of the aggressive act (comp. the last action in Fig. 10.4).

The objective of the abstract condition-action sequences is not to tell the story of a particular case but rather to infer general social mechanisms. For this reason, the reasoning is described in a most general way. Two options had been identified which are characteristic for all cases in the data: to interpret the aggression as norm enforcement or norm violation. Norm enforcement is denoted as 'norm of trust demanded', i.e. as a form of punishment. As the condition of covertness of criminal organisations demands secrecy, it is an advantage not to talk too much. Moreover, norms are not codified. Therefore this is done typically without informing victim that he is being punished because of the violation of a certain norm. Norm violation is denoted as 'norm of trust violated', i.e. as violation of the informal code of conduct within the criminal group (comp. the first branching in Fig. 10.5). Obviously, this broad characterisation covers a number of concrete interpretations. For instance, norm violation might be some kind of self-interested action which can be due to an infinite number of intentions.

Dependent on the interpretation of the aggression different behavioural options are triggered. Obviously the reasoning is not documented in the data. However, what can be found is the reaction on the aggression. First we discuss the case of interpretation as norm enforcement (on the right-hand side of Fig. 10.5). In this case the

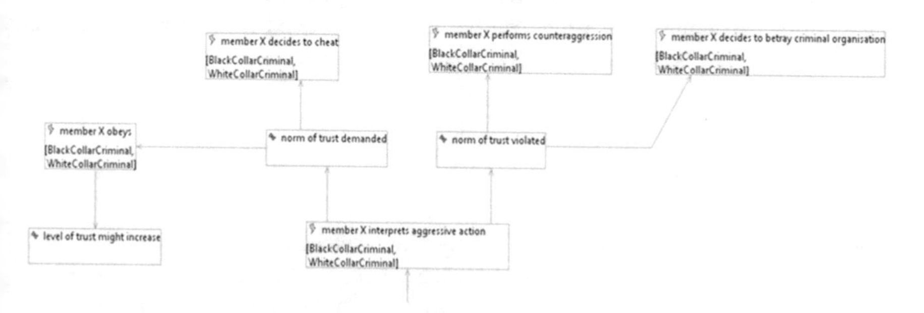

Fig. 10.5 Reasoning on aggression

victim may either obey or deliberately decide to cheat. Obedience may restore the trust in the organisation, or at least ensure that the code of conduct in the group is respected. This is denoted as 'level of trust might increase'. In this case mistrust may fade away or at least remain encapsulated. Obedience is shown as an example of the reaction to the 'request for money':

> I paid but I'm alive.

10.5 Conflict Escalation

In the case of interpreting aggression as norm violation (on the left-hand side of Fig. 10.5), the victim decides about the reaction. Two action classes had been identified, denoted as counter-aggression and betrayal. This shall be discussed by the first example: the failed assassination. This is an intricate case, demonstrating the pathway to the diverging interpretation, 'norm of trust violated'. After M. survived the attack on his life, it is plausible that he lost trust in his business partners. The reaction was as follows:

> M. told the newspapers 'about my role in the network' because he thought that I wanted to kill him to get the money.

This reaction is instructive: It allows reconstructing how he interpreted the aggression. M. interpreted the attack on his life not as a penalty for deviant behaviour from his side (i.e. death penalty as in his later assassination for being accused of stealing drugs). Instead he concluded that the cause of the attack was based on self-interest (the other criminal 'wanted his money'). Thus he interpreted the attack as norm deviation rather than enforcement. Next, he attributed the aggression to an individual person and started a counter-reaction against this particular person by betraying 'his role in the network'. This is an example of betrayal. An example for counter-aggression will be provided when the escalation of the conflicts to a 'corrupt chaos' is discussed. First, it shall be noted that this reaction caused another member of the group to become victim of an act of aggression. While it remains unknown who was responsible for the assassination, it was not this individual. However, the betrayal had severe consequences for this individual. Thus another member of the group faced an act of aggression which further caused the need of interpretation. This induces a positive feedback loop as outlined in Fig. 10.6.

10.5.1 A Corrupt Chaos

Positive feedback loops generate unstable systemic behaviour. This systemic property caused a spreading of mistrust throughout the group. It generated a cycle of revenge and counter-revenge, making the situation uncontrollable, as documented below:

> There is a rule of terror in the town.

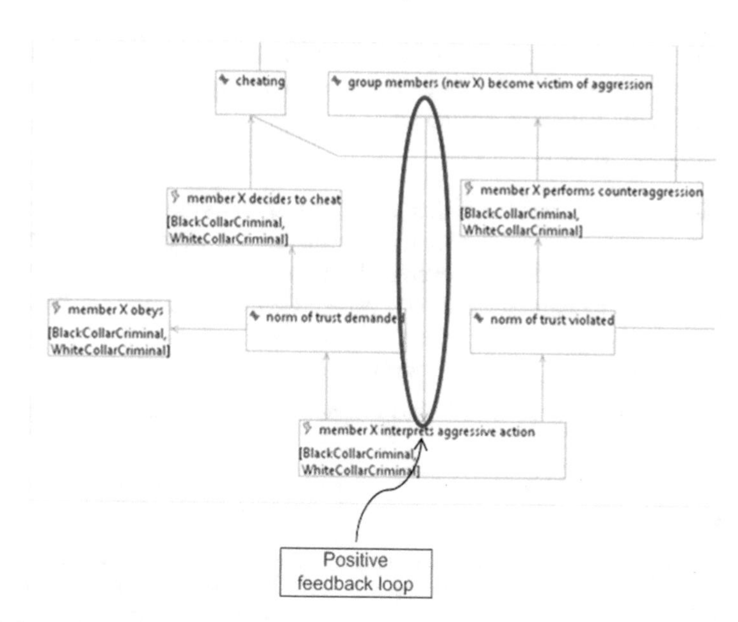

Fig. 10.6 Positive feedback loop

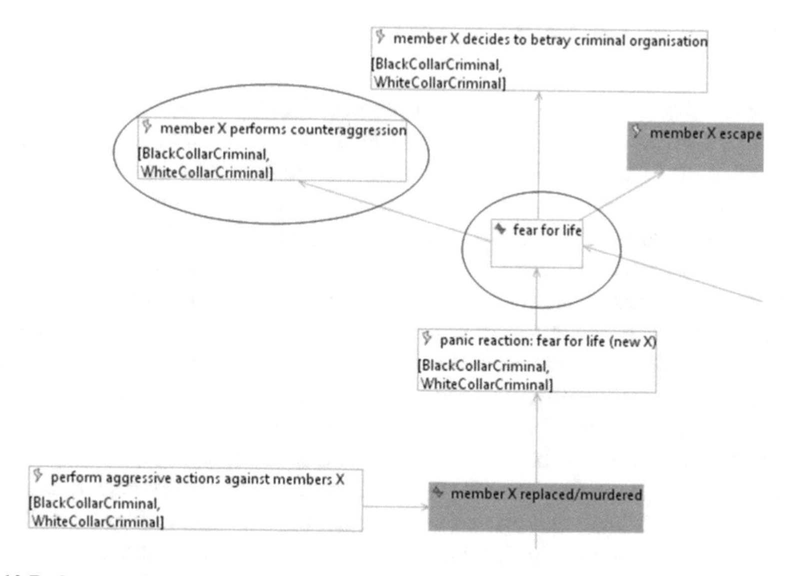

Fig. 10.7 Corrupt chaos

The feedback cycle generates a complexity that from the perspective of the people involved in the situation could not be attributed to a particular individual anymore as indicated in the following statement:

> There is a corrupt chaos behind it.

In the following the condition-action sequences of this segment of the process of disintegration of the group are displayed in which the trust required for the covert activities breaks down. Not the overall diagram will be substantiated by textual annotations from the data. Instead only two elements, denoted as fear for life and counter-aggression, will be highlighted (Fig. 10.7).

Fear for life is proven in the following testimony of a witness:

> V01 was in great fear of O1. When he had an appointment with O1 he was wearing a bulletproof jacket.

However, being thrown in a situation of existential threat is likely to initiate attempts of counter-aggression. This is demonstrated at two examples:

> He was at a point in which he was in a totally despaired situation. HLJ had several times tried to counteract. He had a plan to approach O1 with a weapon. However, in the last moment he didn't dare. At a different time he had two pistols with him. He planned to shoot O1 to death and to pass the other weapon in his hand in order that it appeared as if he had shot in self-defense.
>
> Presumably V01 asked the [Motorcycling gang] to make an operation against O1 in return for a huge amount of money.

10.6 Run on the Bank

Existential threats are likely to induce unpredictable behaviour. However, in the ordinary business of money laundering a huge amount of illegal money had been invested in the legal market through the white collar criminals. In a criminal organisation the investment could not be ensured by legal contracts. The black collar criminals needed to trust that they will get the return of investment back from the white collar criminal. In case of the breakdown of trust a well-known mechanism from legal financial markets becomes effective: fear for money provides an incentive to get as much money of the investment back as soon as possible. Moreover, if it becomes visible that one member attempts to get the money out, the classical mechanism of a self-fulfilling prophecy (Merton, 1968) initiates a 'run on the bank'. It is known from the legal world that this has a destructive effect on the market. In Fig. 10.8 an overview of the process in case of a criminal organisation is provided.

The overall process shown in Fig. 10.8 will now be documented following instantiations of single condition-action sequences. As approved by the following testimony fear for money initiated attempts to get money out of the investment:

> Starting from Oct. XXXX S.K. came in the office. She told the employees that she needed to talk to me because her former man (who died) had 7 million active debts.

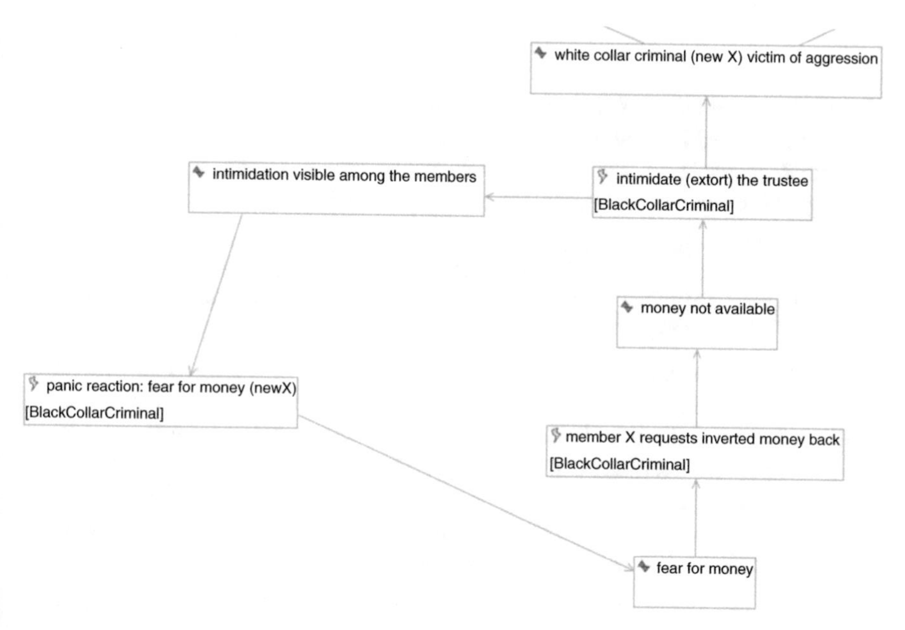

Fig. 10.8 Run on the bank

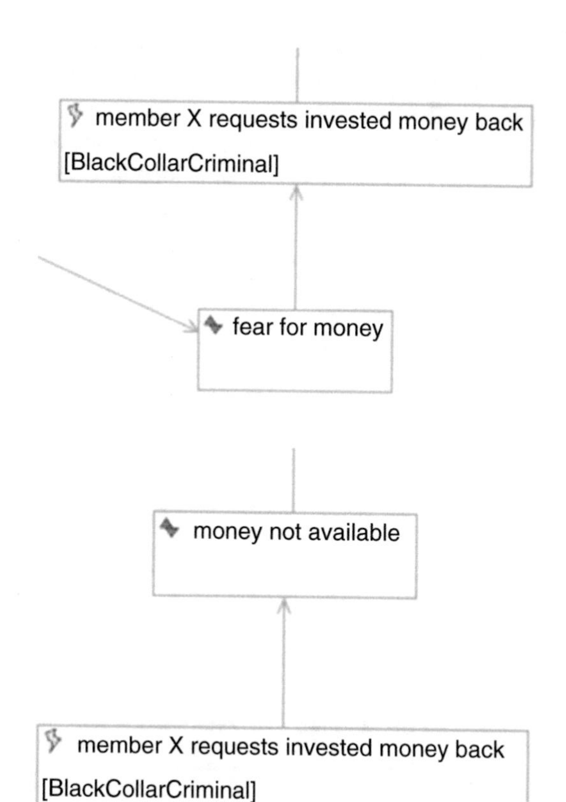

Fig. 10.9 Fear for money

Fig. 10.10 Money not available

This is the beginning of the process as highlighted in Fig. 10.9. However, once money is invested in legal market such as e.g. constructions, the money is no longer immediately available. This is indicated by the condition 'money not available', highlighted in Fig. 10.10, which is testified by several witnesses:

> At a certain point he had problems with his liquidity.
>
> There is a considerable backlog demand in the back-payment. The reason is twofold: first, it's becoming difficult to gain new funding because of the negative reports in the media and second much of our liquidity has been lost in payments to O1.

Since financial claims cannot be enforced by recourse to the court in case of an illegal covert organisation, a run on the bank has the additional effect that the use of violence becomes likely to force the passing over of the money. An attempt to get the money back nevertheless might trigger intimidation of the trustee (the white collar criminal) who now becomes victim of aggression of his business partners. This is highlighted in Fig. 10.11. This results in extortion of the trustee to enforce the claim as testified in two examples below:

> In the last year he was strongly under pressure because he had been extorted. That's what he said to me.
>
> If I don't pay, her Yugoslav friend O6 would kill me.

It is unlikely that intimidation remains secret in the closed community of a small group. Rather rumours might easily spread in the group. This is highlighted in Fig. 10.12. Once attempts to get money out of the investment become visible a new stage of the run on the bank is reached. The business partner might now get 'in fear for money' as well and the same loop as shown in Fig. 10.8 is initiated, now by a new member of the gang. Additional monetary claims generate a cycle of extortion

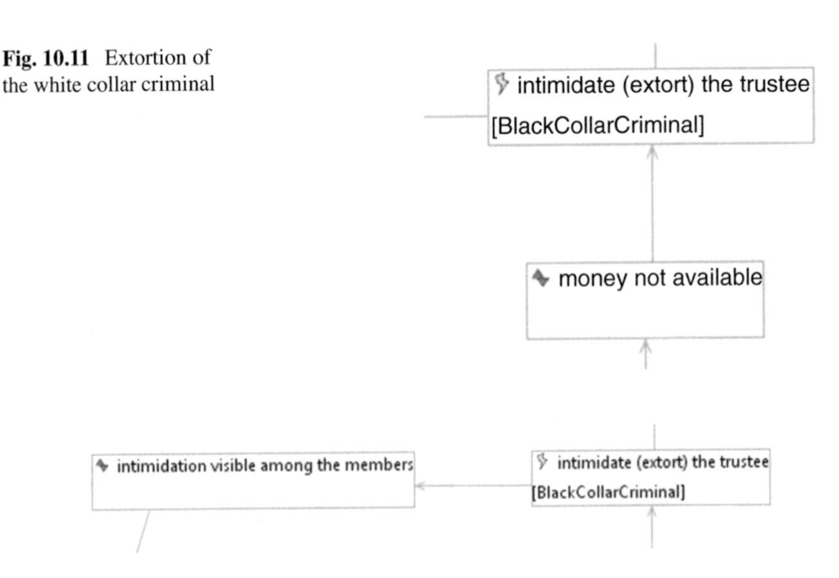

Fig. 10.11 Extortion of the white collar criminal

Fig. 10.12 Positive feedback of fear for money

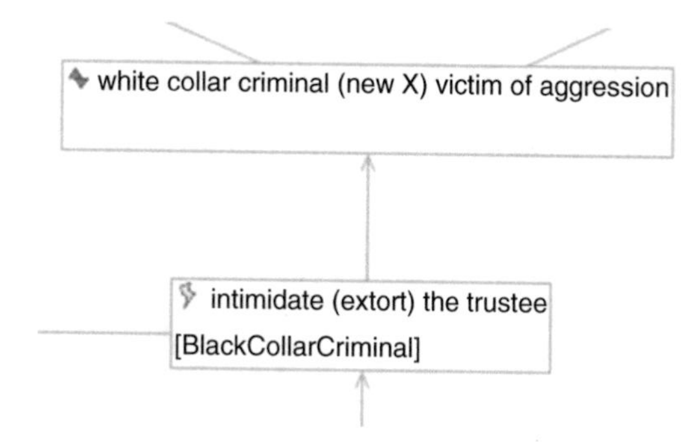

Fig. 10.13 Victimisation of white collar criminal

in order to get the money back. A positive feedback cycle is closed. An example is the second sentence in the following example:

> Soon after his death the widow of K had an affair with O1. She extorted seven million from V01. Contrary to the claim of M. his entitlements had not been captured by this deal.

Thus intimidation stimulates further intimidation making the white collar criminal victim of aggression of his business partners and turning a formerly symbiotic into a parasitic relationship (see Transcrime Joint Research Center on Transnational Crime, 2008). An example of how payment had been enforced is provided below:

> V01 was ordered to the office of his lawyer. However, when he entered the office the lawyer was not there. Instead O1 and seemingly 3 Yugoslavs were there. These ordered him to go on his knees and hold a machine gun in his stomach. (Fig. 10.13)

10.7 Structural Insights of the Conceptual Model

The thick description on the micro level of the *process* of the escalation of violence provides insights in the macro level of the *structural* properties of the group that reveal reasons which triggered the process that finally generated a situation perceived as a 'corrupt chaos'. In abstract terms, the conceptual model describes a cascading effect: Mistrust generated violence which in turn enforced mistrust in the overall group. That such a cascading effect was possible can be ascribed to the organisational structure of the group. Since the group could not rely on formal procedures of conflict regulation, no mechanisms existed to encapsulate the mistrust. This was due to some characteristic features of the group structure which will be described in more detail below.

First it shall be noted that the ordinary business of money laundering reveals a triadic communication structure (see Simmel, 1908), consisting of black collar criminals, white collar criminals and straw men. This is shown in Fig. 10.14. Money is passed from the black collar criminals to the white collar criminals and back via the straw men. In such a situation it is likely that misunderstandings and misperceptions take place. The structure of the relations is a crucial trigger for the spreading of

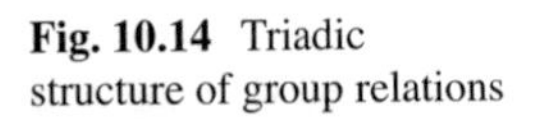

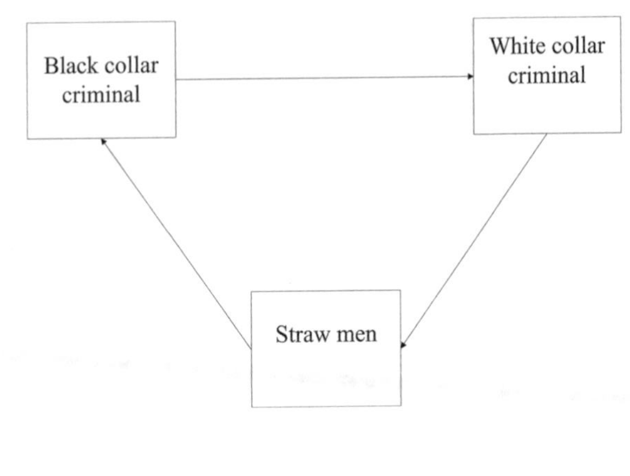

Fig. 10.14 Triadic structure of group relations

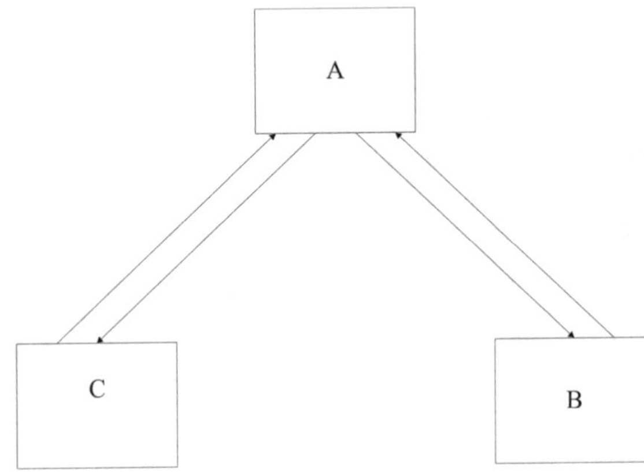

Fig. 10.15 Hierarchic dissolution of triadic relations

mistrust in the group, once initialized. For instance, if money (or drugs for instance) gets lost, the responsibility remains uncertain. The black collar criminal may accuse the white collar criminal or the straw men and both may put the blame on the other. Note that later M. had been killed in such a situation, when he was accused of stealing drugs. Whereas it remains unknown (at least for the 'official world') who was responsible that drugs got lost, execution of a death penalty is an ultimate solution to resolve ambiguity. Thus a triad remains a fragile and unstable structure.

As it is known from Simmel (1908) (see also Sofsky & Paris, 1994), one solution to re-establish stability of social relations is to decompose a triad into a hierarchy. A hierarchy resolves the indecisiveness of the social situation by cutting certain lines of relations. If relations between group members are controlled by a central node in the relational structure (i.e. the top person in the hierarchy), the definition of a situation remains unambiguous. The situation is simply defined by the top person. For instance, the top guy (say A) might convict, e.g. B (see Fig. 10.15) for being guilty of violating the code of conduct. This provides a mechanism of conflict resolution which preserves the secrecy of the group by keeping relations inside the organisation. For this reason Simmel (1908) assumed that covert organisations tend to be hierarchically organised.

However, this is exactly what did *not* happen in this case. Data reveals no formal structures such as a hierarchy or any formal rules of conduct which provide a guideline to handle crisis situations in the group. The structure of the situation was characterised by the following elements:

- While a horizontal differentiation between the tasks of white and black collar criminals existed, on the vertical axis the group had a flat structure. Some informal hubs existed which characterise people involved in many of the actions which had been subject of police investigations. However, while the hubs might have had a certain prestige all members were equal insofar as no individual had a right of command.
- As a consequence of the flat structure of the group, trust was not secured by formal authority but simply based on interpersonal relations. Some individuals knew each other for quite a long time, whereas others such as e.g. straw men had been involved in the activities through a referee. An example is the brother of V01.
- This entails that the norms of conduct remained only implicit.

However, once an initial element of mistrust was intruded, the crisis was characterised by a highly unstructured situation and individuals could not rely on formal rules of crisis management. They had to improvise ad-hoc to react to unanticipated situations such as an attempt of an assassination or reading their names in the newspapers, even betrayed by a criminal comrade. The reaction had to rely on interpretations of the situation. Since the interpretation could not be guided by a formal code of conduct, it remained fallible. Factually the conflict escalation was characterised by misperceptions and diverging interpretations of the situation. The likelihood of such misunderstandings can be traced back to the organisational structure, characterised by a lack of authority which could reduce contingency by providing an unequivocal definition of a situation, simply by its normative power.

10.7.1 Socio-psychological Consequences of the Structural Properties of the Group

As a consequence of the organisational structure the differentiation of punishment and revenge remained blurred. In behavioural terms both actions can be described as an act of aggression. However, both terms constitute social concepts with essential differences with regard to potential follow-up actions: Whereas in case of punishment the aggression might stop once the punishment has been applied, revenge might lead to an endless circle. For instance, if wrong parking is sanctioned by a fee, the violation of the parking norms is compensated once the fee is paid. As endless cycles of blood revenge in traditional cultures demonstrate, the situation is different in case that the social concept of revenge is activated. If somebody becomes victim of aggression it is legitimate and sometimes even prescribed to counter-react with an aggressive act to take revenge. Since in behavioural terms both punishment and revenge is an act of aggression, interpretation is needed to decide about how to react once an individual member of the organisation becomes victim of an aggression. Indeed, the data reveals hints to both interpretations. This is dependent on the subjective perception of the situation. Note that even in case of revenge the individual might be aware that he or she had been subject of aggression because of norm violation. One might recall blood revenge as a paradigmatic example in which people clearly know that they commit a crime which will cause another crime. However, in contrast to an interpretation as punishment the subject does not accept the normative authority of the aggressor.

The identification of the role of a normative authority on the level of the subjective meaning attributed to a certain situation refers back again to structural properties of the organisation. Namely, the validity of norms needs to be secured by a certain form of authority, may this be a formal hierarchy or the authority legitimate reason (Bicchieri, 2006). First and foremost the acceptance of aggressions as legitimate punishment implies that it is possible to identify (legitimate) reasons for the aggression. In a highly unstructured situation as outlined in the escalation of the corrupt chaos the identification for reasons becomes increasingly difficult and fallible. In such a case, namely if recourse to legitimate reasons for aggression become precarious, a formal authority of organisational hierarchies may serve as a substitute of the authority of reason. This role of authority can be illustrated as an example of the relation between parents and their children: At least for children of a certain age who are too young for normative reasoning, it is likely that the child perceives aggression of the parents as a punishment. Even if the parents are wrong or psychopathic alcoholics, children cannot judge if the parents are right or wrong. Conflicts between parents and children are asymmetric. For children parents are the normative power which defines the rules of the world. In other words the family represents a basic social structure. Thus typically aggression is interpreted as punishment. In contrast, in a quarrel between child peers aggression is likely to stimulate revenge since peers are not a normative authority. In abstract terms punishment refers to higher level in a hierarchy or at least to some kind of superior interests. Neither do subordinates punish their boss nor do children punish their parents. While certainly aggression of both children against their parents and subordinates against their boss exists, this is not punishment but rather a violation of the norm of respect to the hierarchy. Thus reasoning on aggression implies reasoning on social structure.

The criminal group had no social structure, except for being a group of peers. Certainly, they were aware of the necessity of their commitment to the group. This was in their interest of making money. The group was based on self-interest and not (as in terrorist groups) on a kind 'moral commitment' based on a certain ideology, may it be Marxism, religion or the nation (comp. also Morselli et al., 2006). Ideologies can provide legitimate reason. This can stimulate a commitment to an abstract higher level authority even in the absence of a formal, real life hierarchical structure. The absence of any formal or ideological authority makes it likely that aggression is countered by revenge rather than interpreting it as legitimate punishment. As organisational science pointed out, a flat social structure has many benefits in terms of reducing transaction costs by providing quick and easy access to information (Williamson, 1981). However, in particular in covert organisations, organisational growth might lead to situations in which interpersonal trust is no longer strong enough to support the organisational structure. Relations of interactions become too complex to be overseen any more. For this reason it becomes increasingly unlikely that possible conflicts could be solved, e.g. by a mediator to which (a) both conflicting parties have a relation of personal trust and (b) which could give a 'wise' judgment that is appreciated by both parties even if this is not backed by a formal authority. Such a mediating process could encapsulate the cascading effect of counter-revenge. However, at a certain point informal conflict regulation becomes unlikely. In consequence organisational growth might reach a tipping point at which a flat structure becomes risky in times of crisis. Even more the structural risk of a triadic structure makes it likely that at some time something goes wrong and crises emerge. Examples from the case are aggression applied to the wrong person. Thus growth calls either for organisational innovation or might entail a risk of organisational failure.

References

Bicchieri, C. (2006). *The grammar of society. The nature and dynamics of social norms*. New York: Cambridge University Press.

Bley, R. (2014). *Rockerkriminalität. Erste empirische Befunde*. Frankfurt/M: Verlag für Polizeiwissenschaft.

Chang, J. J., Lu, H. C., & Chen, M. (2005). Organized crime or individual crime? Endogenous size of a criminal organization and the optimal law enforcement. *Economic Inquiry, 43*(3), 661–675.

Colquitt, J., LePine, J., Piccolo, R., & Zapata, C. (2012). Explaining the justice—Performance relationship: Trust as exchange deepener or trust as uncertainty reducer? *Journal of Applied Psychology, 97*(1), 1–15.

Colquitt, J., & Rodell, J. (2011). Justice, trust, and trustworthiness: A longitudinal analysis integrating three theoretical perspectives. *Academy of Management Journal, 54*(6), 1183–1206.

Corbin, J., & Strauss, A. (2008). *Basics of qualitative research* (3rd ed.). Thousand Oaks: Sage.

Cressey, D. R. (1969). *Theft of the nation: The structure and operations of organized crime in America*. New York: Harper & Row.

Cressey, D. R. (1972). *Criminal organization: Its elementary forms*. New York: Harper & Row.

Diesner, J., Frantz, T., & Carley, K. (2005). Communication networks from the Enron email corpus. It's always about the people. Enron is no different. *Journal of Computational and Mathematical Organization Theory, 11*(3), 201–228.

Gottschalk, P. (2010). Criminal entrepreneurial behavior. *Journal of international business and entrepreneurship development, 5*(1), 63–76.

Gross, E. (1978). Organizational crime: A theoretical perspective. *Studies in Symbolic Interaction, 1*, 55–85.

Hedström, P., & Ylkoski, P. (2010). Causal mechanisms in the social sciences. *Annual Review of Sociology, 36*, 49–67.

Hobbs, D. (2001). The firm: Organizational logic and criminal culture on a shifting terrain. *British Journal of Criminology, 41*(4), 549–560.

Klerks, P. (2002). The network paradigm applied to criminal organizations. *Connections, 24*(3), 53–65.
Krebs, V. (2002). Mapping networks of terrorist cells. *Connections, 24*(3), 43–52.
La Spina, A. (2005). *Mafia, legalità debole e sviluppo del Mezzogiorno*. Bologna: il Mulino.
Lotzmann, U., Neumann, M., & Moehring, M. (2015). From text to agents: Process of developing evidence based simulation models. In *Proceedings of the 29th European Conference on Modelling and Simulation*.
Lotzmann, U., & Wimmer, M. (2013). Traceability in evidence-based policy simulation. In *Proceedings of the 27th European Conference on Modelling and Simulation, ECMS 2013*. Dudweiler: Digitaldruck Pirrot GmbH, pp. 696–702.
Merton, R. (1968). *Social theory and social structure*. New York: Free Press.
Morselli, C. (2009). *Inside criminal networks*. New York: Springer.
Morselli, C., Giguere, C., & Petit, K. (2006). The efficiency/security trade-off in criminal networks. *Social Networks, 29*(1), 143–153.
Natarajan, M. (2006). Understanding the structure of a large heroin trafficking network: A quantitative analysis of qualitative data. *Journal of Quantitative Criminology, 22*(2), 171–192.
Neumann, M., & Lotzmann, U. (2014). Modelling the collapse of a criminal network. In *Proceedings of the 28th European Conference on Modelling and Simulation, ECMS 2014*. Brescia.
Neumann, M., & Sartor, N. (2016). A semantic network analysis of laundering drug money. *Journal of Tax Administration* (forthcoming).
Paoli, L. (2003). *Mafia Brotherhoods. Organized crime, Italian style*. Oxford: Oxford University Press.
Putten, C. van (2012). *The process of extortion: Problems and qualifications*. In Conference on Extortion Racket Systems. University of Vienna, Vienna, pp. 7—11.
Reuter, P. (1983). *Disorganized crime: The economics of the visible hand*. Cambridge, MA: MIT Press.
Scaglione, A. (2011). *Reti Mafiose. Cosa Nostra e Camorra: organizzazioni criminali a confronto*. Milano: FrancoAngeli.
Scherer, S., Wimmer, M., Lotzmann, U., Moss, S., & Pinotti, D. (2015). An evidence-based and conceptual model-driven approach for agent-based policy modelling. *Journal of Artificial Societies and Social Simulation, 18*(3), 14.
Scherer, S., Wimmer, M., & Markisic, S. (2013). Bridging narrative scenario texts and formal policy modelling through conceptual policy modelling. *Artificial intelligence and law, 21*(4), 455–484.
Simmel, G. (1908). *Soziologie. Untersuchung über die Formen der Vergesellschaftung*. Berlin: Duncker & Humblot.
Sofsky, W. (1996). *Traktat über Gewalt*. Frankfurt/M: Fischer.
Sofsky, W., & Paris, R. (1994). *Figurationen sozialer Macht*. Frankfurt a.M: Surkamp.
Sparrow, M. (1991). The application of network analysis to criminal intelligence: An assessment of the prospects. *Social Networks, 13*(3), 251–274.
Transcrime Joint Research Center on Transnational Crime (2008). *Study on extortion racketeering. The need for an instrument to combat activities of organized crime*. Final report. University degli studi di Trento and Universita Cattolica del Sacro Cuore di Milano.
von Lampe, K. (2015). Big business: Scale of operation, organizational size, and the level of integration into the legal economy as key parameters for understanding the development of illegal enterprises. *Trends in Organized Crime, 18*(4), 289–310.
Weick, K. (2007). *Der Prozess des Organisierens*. Frankfurt a. M: Suhrkamp.
Williamson, O. (1981). The economics of organizations. The transaction cost approach. *American Journal of Sociology, 87*(3), 548–577.

Chapter 11
A Simulation Model of Intra-organisational Conflict Regulation in the Crime World

Ulf Lotzmann and Martin Neumann

11.1 Introduction

This chapter can be seen as a sequel to Chap. 10, where the qualitative analysis of texts from police investigations about a criminal network was transformed into a conceptual model of the dynamics that led to the violent breakdown of this network. This conceptual model was transformed and formalised into a simulation model, which is described in this chapter. Subsequently, results gained from simulation experiments with the model are presented.

11.2 Simulation Model Description

The implementation of the simulation model follows closely the modelling process developed in the EU project OCOPOMO,[1] and uses the toolbox provided by this project. The conceptual model was developed with the CCD Tool—the core component of the OCOPOMO Toolbox—which also provides a transformation tool called CCD2DRAMS that allows the semi-automatic transformation into a basic simulation model. The applied modelling process is presented in (Lotzmann, Neumann, & Möhring, 2015). The target platform of this transformation tool is the popular simulation framework Repast (North, Collier, & Vos, 2006), with the declarative rule engine DRAMS (Lotzmann & Meyer, 2011) as an extension for specifying the agent behaviour. Primarily the use of DRAMS shapes the implementation style in a

[1] http://www.ocopomo.eu/.

U. Lotzmann (✉) • M. Neumann, Ph.D.
Department of Computer Science, Institute for Information Systems Research, University of Koblenz-Landau, Universitätsstr 1, Koblenz 56070, Germany
e-mail: ulf@uni-koblenz.de; maneumann@uni-koblenz.de

C. Elsenbroich et al. (eds.), *Social Dimensions of Organised Crime*, Computational Social Sciences, DOI 10.1007/978-3-319-45169-5_11

particular direction: The entire agent behaviour is specified by declarative rules, which operate on the knowledge stored as facts in the so-called fact bases. As DRAMS is designed as a distributed rule engine, each agent is equipped with its own fact base and own rules, while for 'world knowledge' and also communication purposes a global fact base is provided. Also global rules are allowed to implement activities that cannot be located to concrete agents. Each rule consists of a condition part, the so-called left-hand side (LHS) and an action part, the right-hand side (RHS). The conditions in the LHS are specified using a set of clauses, e.g. for performing fact base queries, binding variables, comparing variables and constants, doing mathematical calculations and so on. The RHS consists of clauses that allow for modifications of fact bases (asserting new facts, retracting existing facts) as well as clauses for writing simulation outcomes in different ways. The basic mechanism of the rule engine is then to evaluate the LHS of all rules for which the facts are available and other matching conditions are fulfilled, and then fire the rule by executing the RHS, setting the condition for new rules to fire, and generating the simulation log.

The actual implementation of the simulation model follows closely the conceptual model, not least due to the code generation facility provided by the toolbox. All the actions modelled in the CCD action diagram are also present as DRAMS rules in the simulation model. In order to achieve a consistent implementation, a number of aspects had to be added to the model which are not described in the evidence base, instead relying on cognitive heuristics. On the other hand, some details included in the conceptual model had to be left out to keep the complexity of the simulation model manageable, but also due to decisions to concentrate the focus on some crucial aspects of interest for deeper analysis of the case. These implementation decisions were in most instances discussed with the data analysis expert and partly also with domain experts.

Another reason to ground the simulation model on DRAMS is the opportunity to benefit from the traceability functionality built in the OCOPOMO toolbox (Lotzmann & Wimmer, 2013). Herewith it becomes possible to trace simulation results back to the phrases from the evidence base annotated to elements of the conceptual model. That is, this functionality opens a way to efficiently perform qualitative analysis of simulation results by means of unveiling the relations between dynamics in simulations runs and events in the real criminal network described in the evidence base.

The following section gives an overview of the simulation model both in terms of static and dynamic aspects. The former includes the agents and related attributes from which the model is comprised, the latter the control flow in the different parts of the model. In the subsequent sections this control flow is further detailed in order to give quite deep insights on concrete design decisions to show how the evidence is reflected in the implementation.

11.3 Simulation Model Overview

In the simulation model agents are included for the CCD actor types Black Collar Criminal, White Collar Criminal and Police (cf. Chap. 10 on white collar and black collar criminals). While for the two types of criminals arbitrary numbers of instances can be set for simulation runs, the police is represented as an institutional agent, i.e. a single agent instance covers the activities of this actor. In a typical simulation run there exists a single White Collar Criminal, who is responsible for money laundering and is typically also part of the legal world, but might become involved in aggressive practices of the Black Collar Criminals. These are the actual representatives of the illegal world of the criminal network. There are two types of Black Collars distinguished, one called the Reputable Criminal which is initially in the so-called rational mental frame, while the other 'ordinary' Criminal only acts in the emotional mental frame. In the course of the simulation also the Reputable Criminal might switch to the emotional frame, e.g. due to violent events. This distinction between the two types of mental frame is illustrated below and argued in Chap. 10.

The model is implemented in a tick-based way where the course of time is represented by discrete ticks, but no defined time period between ticks is specified. Actions or reactions involving multi-staged decisions process are typically spread across a number of ticks, as are the consequences of actions and police investigations, to give a few examples.

This temporal relationship is one of the pieces of information given in the activity diagram in Fig. 11.1, which furthermore shows the control flow between the important behavioural elements (represented by activities) of the entire model, structured in different parts (grey background boxes). Some of the edges are labelled in order to improve readability. So are temporal relations as mentioned above put in square brackets, phrases in italics give further details on conditions, if the subsequent activities do not allow inferring this information. The most important edge label is printed in bold font: the (type of) agent who is executor of the following activity. Edges with no label indicate the transition to the next activity within the same tick and as part of the behaviour of the same agent. The diagram can be read as follows.

The dynamics start with an initial normative event at the first tick regarding a random criminal. This is an unspecified violation of intra-organisational norms that stimulates the necessity of conflict regulation. This normative event is observed by fellow criminal at the next tick, who might adapt the image of this criminal. In case of a norm violation event the image is decreased, which triggers a decision process on whether and how to perform aggressive actions against the deviating criminal. In the next tick the possibly many criminals who decided to sanction the norm violation 'negotiate', and finally one of them performs a single aggression, whose consequence manifests in the next tick:

Either the aggression is lethal, which might cause panic and 'fear for life' among other members of the criminal network, or the victim of the aggression experiences the aggression and starts with an interpretation process.

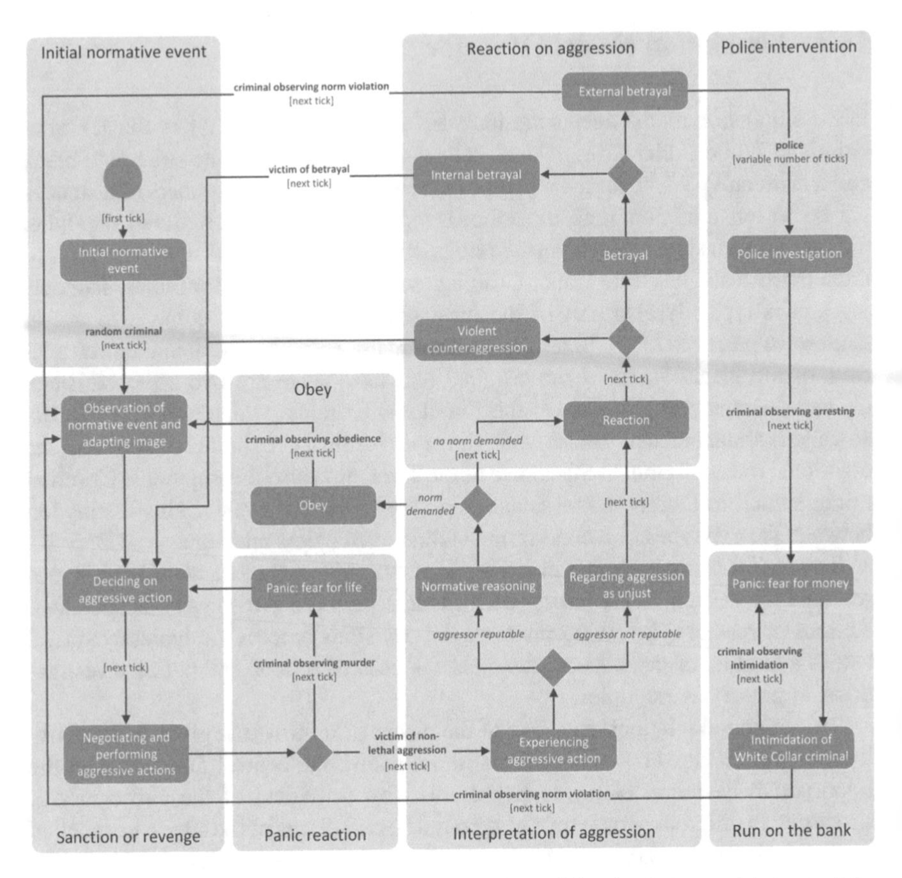

Fig. 11.1 UML activity diagram providing an overview of the simulation model (see text for meaning of edge label styles)

This interpretation begins with the distinction whether the aggressor is reputable or not. In the latter case, the aggression is regarded as unjust which triggers an obligatory reaction in the next tick. If the aggressor is judged to be reputable, then a normative process is performed that leads to the conclusion that either a norm is indeed demanded, persuading the criminal to obey to the norm (which in the next tick might motivate fellow criminals to increase the image of this member, if they get to know about the obedience), or no norm is demanded which again triggers an aggressive reaction in the next tick.

About the actual reaction a decision process is conducted (taking one more tick), with one of the following results:

- A violent counteraggression is performed, employing the same activities as for normative sanctioning (as described above), this time of course executed by the reacting agent.

- The criminal that issued the original aggression is betrayed internally, i.e. involving just the two criminals. The victim of this betrayal will decide on a responding aggression in the next tick.
- An external betrayal is performed, which can either be to inform the police or to go to the media and revealing the criminal network (or its members) to the public. Both actions trigger police investigations, while the latter one in addition is recognised as a norm violation, which might be observed by fellow criminals in the next tick and might furthermore lead to the already known consequences of new aggressive actions.

Police investigations ultimately lead to interventions, i.e. the arresting of members of the network. This arresting might also be observed by other members and in the next tick cause a panic about the potential loss of invested money. This fear usually triggers an intimidation of the White Collar Criminal, which might also be observed by other criminals, starting (with a time delay of one tick) a vicious cycle of cascading acts of extortion towards the White Collar in form of a 'run on the bank'. The refusal of repayment of invested money by the White Collar is at the same time regarded as a norm violation, observable by further criminals (again with a delay of one tick).

11.4 Decision Processes

The functional blocks shown in Fig. 11.1 are described in more detail in the following subsections, complementing the very brief walk through the model. A number of concepts partly introduced already are repeatedly used throughout the chapter. These are as follows:

- Rational and emotional mental frame. As mentioned above, these different 'modes of operation' of criminals influence their behaviour. In the emotional frame the criminal is less able to foresee the consequences of the performed actions; hence, the probability for severe aggressions and acts of strong violence is higher than for the rationally acting criminal (cf. Chap. 10).
- Following Sabater-Mir, Paolucci, and Conte (2006) criminals are endowed with image and reputation. Both are properties expressing the standing of a criminal, the rank in the hierarchy in a way. Reputation is initially set for each criminal agent in the initialisation of a simulation run, is known to all members of the criminal network and does not change in the course of time. In contrast, the image is an information private to each criminal agent. That is, each criminal has his own view on the image of each fellow criminal. The image values do change during simulation runs. Thus reputation is an objective property of the criminals while image denotes the subjective evaluation of the fellow criminals by each member of the gang.
- Levels of image and reputation. These are ordinal scaled attributes: very high, high, modest low and very low.
- Levels of severity of aggressive actions. The severity of an aggressive action is measured by the ordinal scaled attribute 'strength': low, modest and high.

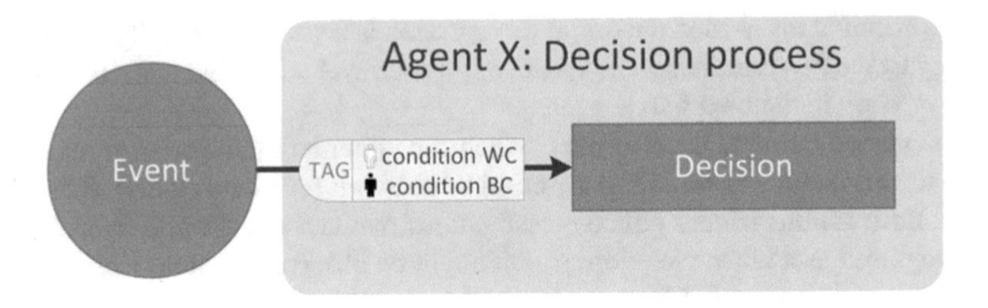

Fig. 11.2 Notation of decision trees used in the chapter

Some of the detailed descriptions in the subsequent sections use decision trees to illustrate the model behaviour. Fig. 11.2 shows an example of such a tree. The circle represents the trigger event, starting the decision process. The different stages of the process are displayed as rectangles, while the specific condition for a decision is attached to the respective edge. If the conditions differentiate for the different kinds of criminal agents, a little black or white actor symbol is shown, referring to the Black Collar and White Collar Criminal, respectively. Some of the edge descriptions are extended by a tag, which refers to a concrete description in the text. These are typically examples for important parameters of the simulation model.

11.4.1 Initial Normative Event

To create the initial event that a member of the criminal network all of a sudden becomes disreputable—as discussed in Chap. 10—a global rule throwing an external event is provided. This rule picks randomly one of the members and issues a normative event about an alleged violation of the norm of trust by this member. This event is triggered just once, at tick 1.0.

11.4.2 Sanction or Revenge

This functional block basically implements the CCD action 'perform aggressive actions against member X' (Fig. 11.3), and is an example where the implementation that formalises this action is much more convoluted than the action might indicate. Reason for this discrepancy in granularity is the fact that for this action not much evidence is available—the internal decision processes of criminals that lead to aggressive actions have to be regarded as a black box. Therefore, the mechanisms have to be constructed in some plausible and—where possible well informed—way, with the provisio that the events known from empirical evidence can be observed in simulations as results of these decisions.

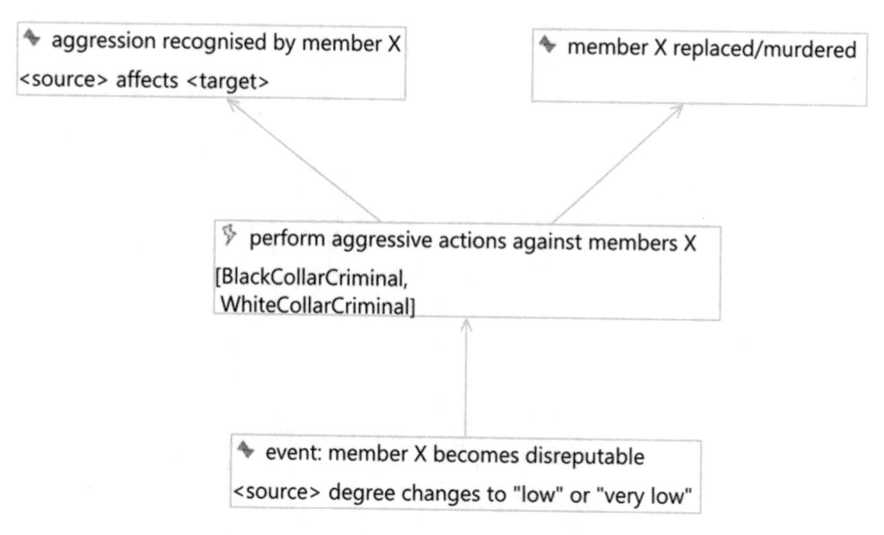

Fig. 11.3 CCD action 'perform aggressive actions against member X'

In Fig. 11.4 the decision tree formalising this action is shown. The initial condition—a criminal 'X' violates a norm—can in principle be observed by each fellow criminal and might lead to a reaction. This perception involves the observation of the event, but also the 'willingness' to care about the event, and is modelled as a stochastic process.

As annotated in (A1), the White Collar criminal perceives this event with a very low probability of 0.05 since he typically keeps out of the thuggish business of the Black Collar criminals, while for the Black Collar criminal the probability is dependent on the image of the criminal respective to the event: with a very high image, the probability is 0.1, with high image 0.2 and otherwise 0.3. The rationale behind this differentiation is that a norm deviation of a criminal with higher image seems less likely to be an offending act against fellow criminals or a threat for the entire network.

The first step of the process that is triggered on successful perception is a change in the image of the criminal. If the normative action was a norm violation, then the image strongly decreases ('two levels'); in the case of norm obedience (not shown in the decision tree) the image increases by one 'level'.

The new image of the criminal related to the normative event then triggers the next step of the decision process, where the behaviour differs if the criminal is in rational or emotional mental frame.

In both cases an aggression is planned only if the new image of the criminal related to the normative event is low or even very low, but in the case of rational frame the planning is followed only with a probability of 0.9 (A2), whereas an emotionally acting criminal would always punish, because he might not be able to foresee the consequences of his aggressive actions (A3).

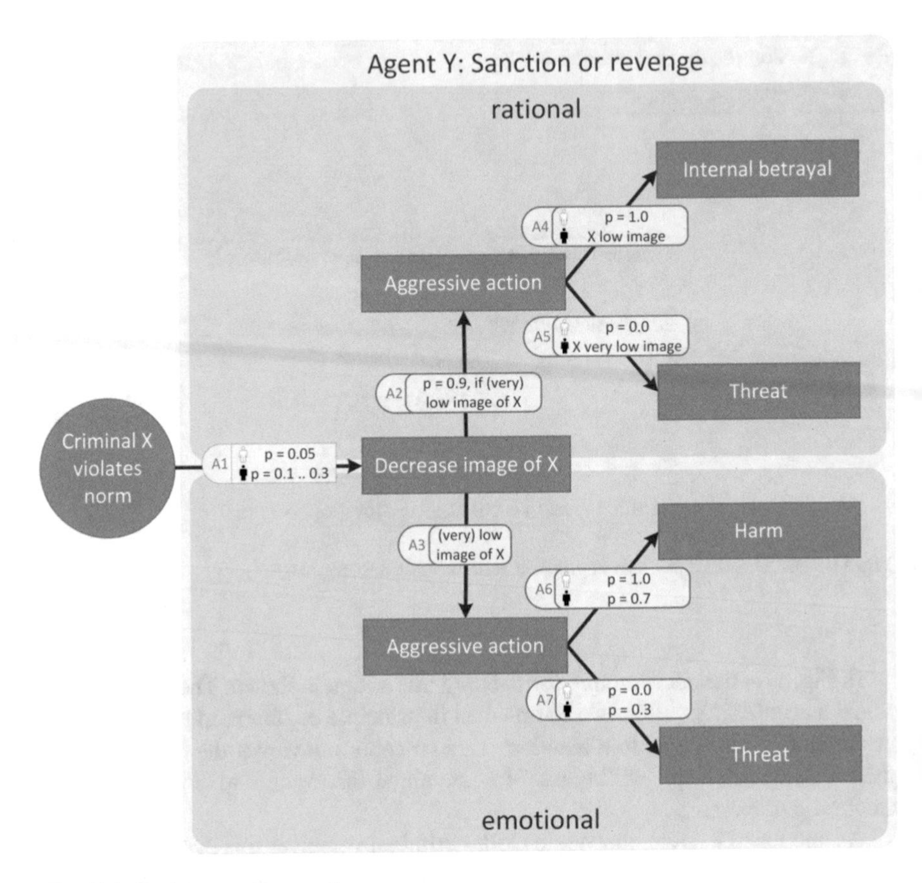

Fig. 11.4 Decision tree for sanction or revenge

If once the plan is conceived, then again the category of criminal (Black or White Collar) and the mental frame determine the type of reactions, but in some cases also the image of the criminal to be punished (decisions A4 and A5 of Black Collar criminal).

A rational White Collar criminal will always (A4) perform the—compared to the other options—mild punishment of (internal) betrayal, while in the emotional frame he will always answer with violence (A6). A rational Black Collar criminal considers the option of betrayal only if the target of the aggression has still a low image (A4); in the case of very low image (A5), the only appropriate action is considered to be threat. An emotionally acting Black Collar criminal tends more towards a violent reaction (probability of 0.7; A6) than a threatening action (probability of 0.3; A7). The cognitive heuristics modelled in these decisions are suggested by information from the evidence base; this connection to evidence becomes more concrete when deciding on the actual aggressive action. Figure 11.5 shows this decision process for internal betrayal, Fig. 11.6 for threat and Fig. 11.7 for harm. The

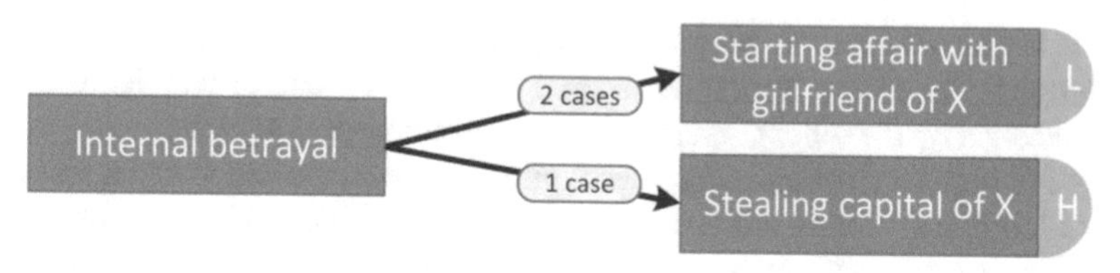

Fig. 11.5 Decision tree for internal betrayal

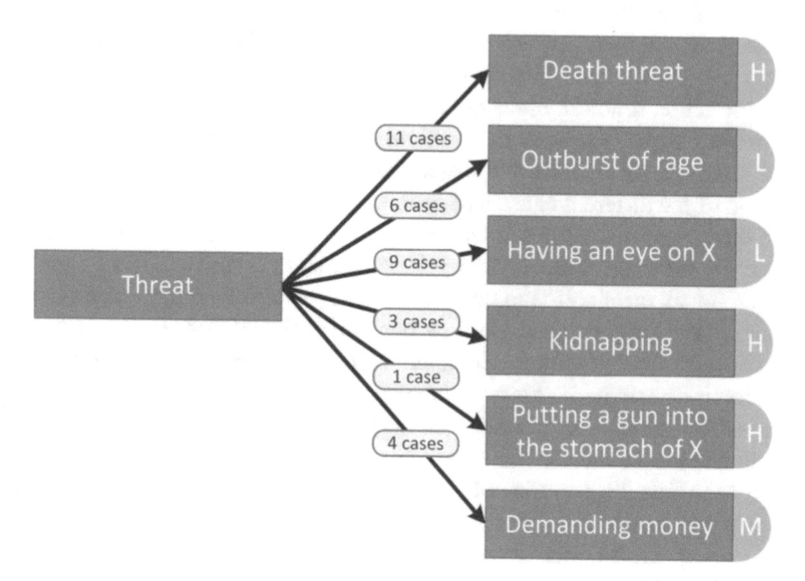

Fig. 11.6 Decision tree for threatening actions

Fig. 11.7 Decision tree for violent actions

letters right to the decision boxes point to the severity of the action: (H)igh, (M)odest and (L)ow.

As a side remark, the decision trees in Fig. 11.5 to Fig. 11.7 and also in Fig. 11.16 refer back to the first phase of a qualitative analysis as described in Chap. 10. In the step of the analysis of the textual data the so-called in vivo codes had been created, i.e. annotations of characteristic brief text elements. These had then been subsumed to broader categories which provide the building blocks of the conceptual model. In Chap. 10 we only provided examples of these in vivo codes to illustrate how the categories can be traced back to empirical evidence. However, the relative frequency of in vivo codes subsumed to the different acts of betrayal, threatening or violence enables a specification of the probabilities. Certainly these have to be used with caution: first the categorisation has been undertaken by one of us (Neumann) and cross-checked by the other (Lotzmann). Thus an element of subjective arbitrariness comes into play when subsuming a description of a concrete action under a category such as 'outburst of rage' etc. Second, the relative frequencies in data might not be very reliable. As they are based on police interrogations, an event such as an attempted assassination is more likely to be subject of the interrogation than, e.g. an 'outburst of rage'. It might well be the case that the respondents did not remember or that the talk simply did not approach the issue. Nevertheless, for instance the high absolute number of death threats or attempted assassinations compared to other courses of action found in the data provides a hint for the high disposition of violence in the group. Thus given the problem of dark figures inherent in any criminological research, the relative frequencies provide at least a hint of the empirical likelihood of the different courses of action.

The finally decided aggression is then 'discussed' among the agents that decided to react. So the final result of this decision process is an individual aggression. However, not all criminals who decided to react perform their aggression individually, but rather a single aggression is perpetrated against the criminal X. After some kind of 'negotiation' among the potential aggressors, one aggressor and the related aggression is determined. This two-staged process is not associated with any agent, but part of the 'global' environment. In the first stage the criminal with the highest image is chosen. In the second stage the aggression with the highest severity (as referred to in Fig. 11.5 to Fig. 11.7) is selected, involving a stochastic process if more than one candidate fulfils these criteria.

The subsequent (implicit) execution of the aggression is immediately evaluated in terms of impact for the victim (by another global rule). For acts of violence a certain (quite low) probability for lethal consequences are considered (0.2 for murder attempt, 0.1 for beating-up). All other possible types of aggression are not assumed to be lethal, anyway.

11.4.3 Panic Reaction

If the aggression turned out to be lethal, the CCD action for 'fear for life' panic in Fig. 11.8 comes into play. This is a simplification of the original action diagram from Chap. 10 where also a more general panic might lead to fear for life, if the overall trust in the criminal network has been destroyed. This simplified implementation, however, covers the aspect of loss of trust in the network quite well, based on the loss of image of individual fellow criminals.

Panic is a situation where rational deliberations do no longer play a role in the behaviour of the individual criminal. It plays a central role in terms of escalating aggression and violence among the network members.

However, the implementation as shown in Fig. 11.9 does not need to be particularly complicated. As soon as a murder of a fellow criminal is observed, the 'fear for

Fig. 11.8 CCD action 'panic reaction: fear for life (new X)'

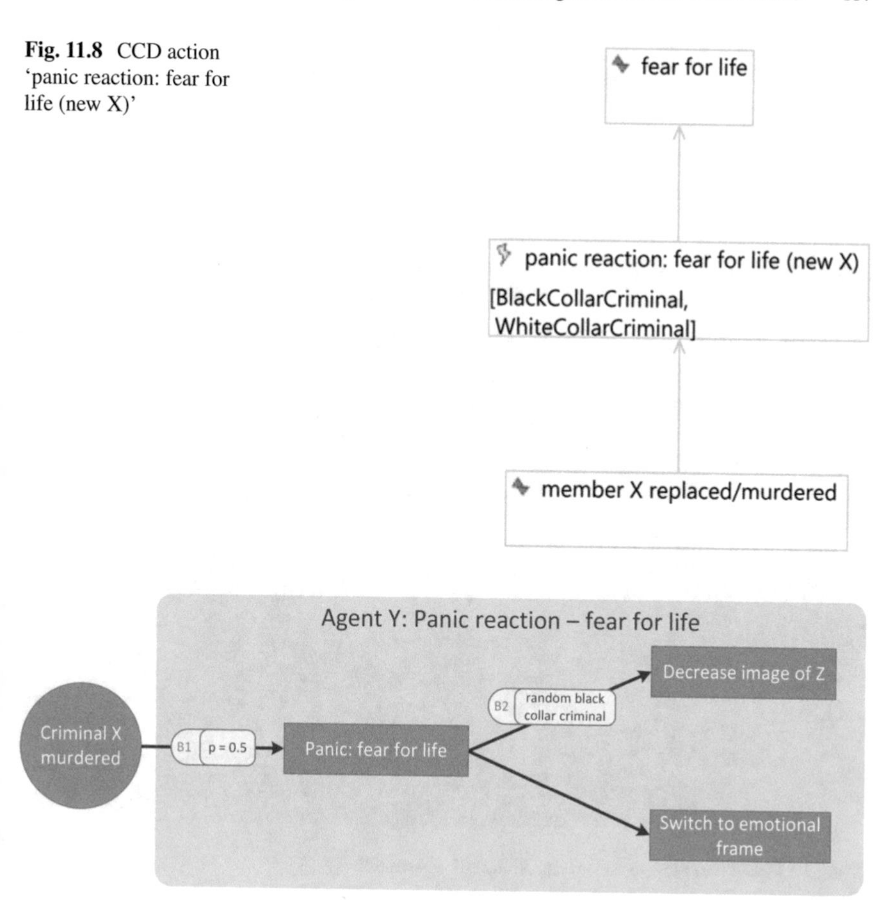

Fig. 11.9 Decision tree for panic reaction: fear for life

life' panic state is established with a probability of 0.5 (B1). The criminal 'switches' into emotional frame and becomes active in some way in order to defend himself. Here he just picks randomly one of the fellow criminals (B2)—which can be interpreted as the one guilty for the murder as perceived by the criminal in panic—and just decreases strongly the related image. This decrease of image might then trigger further actions, as shown in Fig. 11.9. Hence, the spiral of violence escalates.

11.4.4 Interpretation of Aggression

If the aggression was not lethal, the victim has to interpret the reasons for being attacked.

This is modelled in the part of the CCD action diagram displayed in Fig. 11.10. In addition to the condition focussed on here—the aggression motivated by an

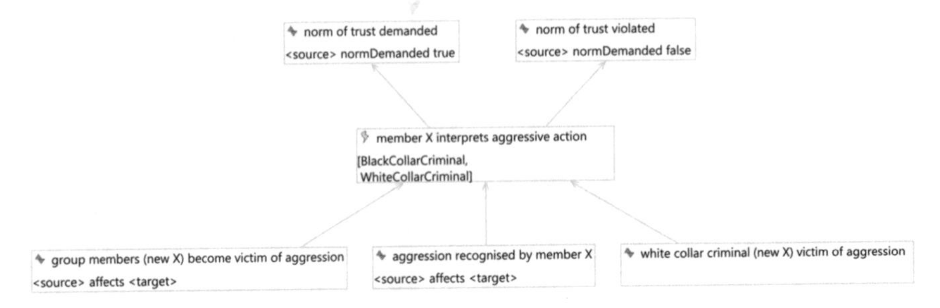

Fig. 11.10 CCD action 'member X interprets aggressive action'

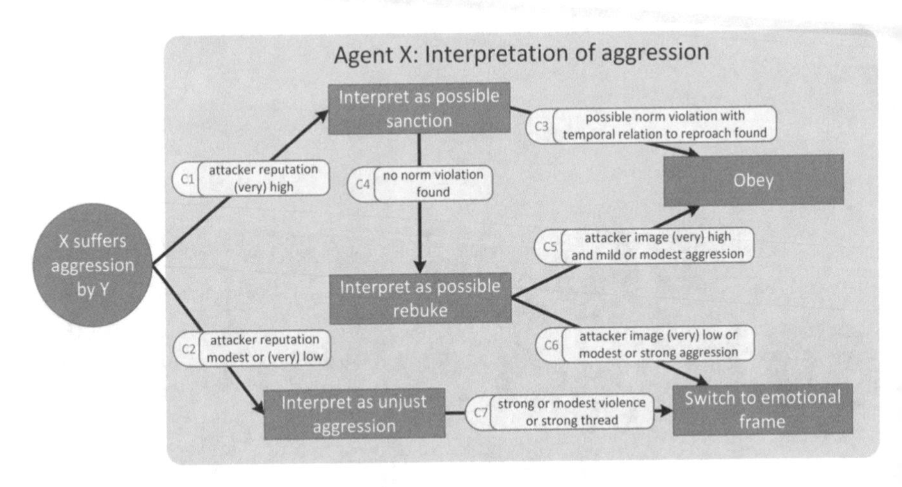

Fig. 11.11 Decision graph for interpretation of aggression

alleged norm deviation 'recognised by member X'—there are two other circumstances when a criminal becomes victim of an aggression: either as a result of a counteraggression or—as a special case—the intimidation of the White Collar criminal. Both cases are not directly linked to a normative event and become relevant at later stages in the dynamics. This interpretation process remains the same for all cases.

In the implementation it is assumed that the victim always perceives the aggression against him. In the following, the implementation of the quite complicated reasoning process is spread out in more detail.

This interpretation process, again, consists of several sub-processes, as shown in Fig. 11.11. The first stage covers the perception of the aggression, followed by a first evaluation of the appearance of the attacker—deduced from the attacker's reputation.

Dependent on this reputation information, the interpretation is fundamentally different. For the case of a reputable attacker (C1) the second stage is a reasoning about whether the attacked criminal might have violated a norm in the recent past which would have led to a sanction of another fellow criminal. This interpretation

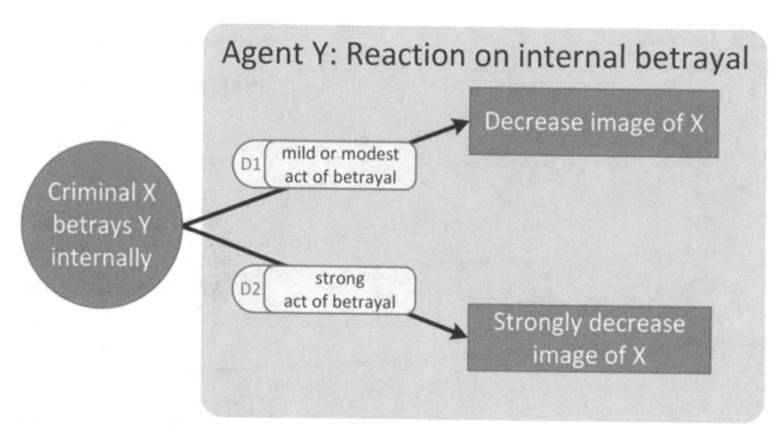

Fig. 11.12 Decision tree for the reaction on internal betrayal

as a possible sanction is done by a 'normative process'. The basic idea of this normative reasoning is quite simple: It is evaluated whether own aggressive actions performed in the past stand in some kind of temporal relationship with a normative event assigned to this criminal. In order to conduct this evaluation, each criminal can access a global event board where all aggressions performed by each criminal are recorded. Also the normative events are logged in a similar way, so that temporal relations between these types of events can easily be derived. The normative process is considered successful, if aggressions are found which at most 16 ticks later led to normative events (C3). If such relations exist, the criminal regards a norm demanded and typically reacts by obeying to the normative request.

Even if the normative process failed (C4), the aggression might still be regarded as a justified sanction: If the attacker has a high or very high image and the aggression was mild or modest (C5), then it is assumed that a norm is demanded as well. This cognitive heuristic has been included in the model to cover the possible aptitude of criminals with high image (and high reputation) to mitigate conflicts, either by mediating or by just exercising authority.

In contrast, if either the attacker's image is modest or low (C2), or the aggression was of high severity ('strong aggression'; C6), then the aggression is perceived as arbitrary, which means that no norm can be demanded. As victim of such a kind of aggression the change into the emotional mental frame appears to be indicated. The same holds for the case of a non-reputable attacker: the aggression is interpreted as unjust. As a consequence the mental frame might change to emotional (due to fear or rage), namely in case of strong or modest violence or strong threat (C7). Entering the emotional frame triggers a reaction, as described in the next but one subsection.

A special case is the internal betrayal, where the affected criminal always reacts by decreasing the image of the betraying criminal (Fig. 11.12). A betrayal with low or modest severity initiates a decrease of image by one level (D1), while highly severe betrayal causes a drop by two levels (D2). Without bothering about

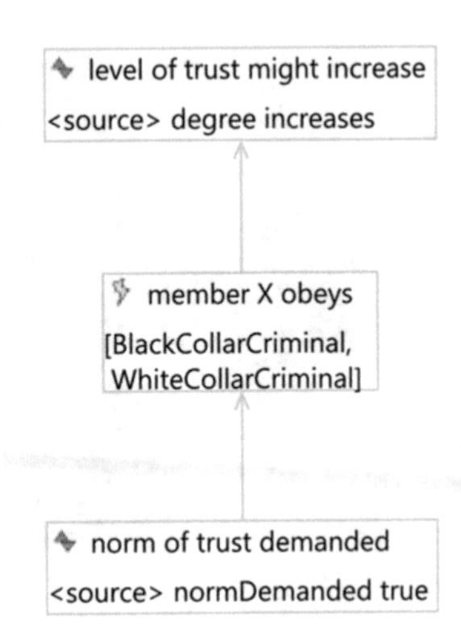

Fig. 11.13 CCD action 'member X obeys'

normative reasoning, a responding action might immediately be triggered according to Fig. 11.4.

11.4.5 Obey

The action diagram fragment shown in Fig. 11.13 covers the criminal's reaction if the normative reasoning resulted in the insight that the experienced aggression is likely to be justified, be it as a sanction for an own actual norm violation or just due to the high image of the aggressor. The only possible action implemented here is to obey the aggression in order to recover the trust among the criminals.

The respective implementation in the simulation model foresees two possibilities for this obeying behaviour, one for each of the two circumstances to obey as mentioned above:

- a rule 'member X obeys', if the normative process classified the aggression as a sanction, and
- a rule 'member X obeys due to high image of aggressor'.

The result in both cases is the same: a normative event carrying the message that the criminal is willing to obey to the norm of trust is send to the environment, i.e. made known to the other members of the criminal network, who might react by increasing the image of the obeying member.

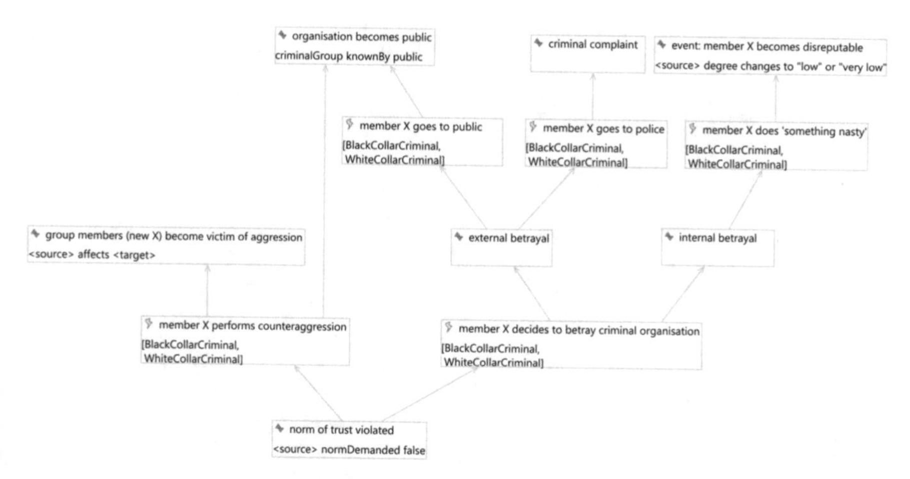

Fig. 11.14 CCD action diagram for reaction on aggression

11.4.6 Reaction on Aggression

The opposite result of the normative reasoning is the awareness that the aggression cannot be a justified sanction or the aggressor has such a low image that he is ineligible to be a sanctioner (i.e. aggression by non-reputable attacker). This particular instance only leaves margin for two types of reaction, either betrayal or violent aggression. This is conceptually modelled in the CCD action diagram fragment shown in Fig. 11.14.

The pre-selection between betrayal and violence is implicitly modelled in the different actions branching from the condition 'norm of trust violated'. One of the two options to respond to unjust aggression is to perform counteraggression, which in this context means some kind of violent act. The second option is to betray the criminal network. There are basically two possible categories of betrayal: the quite harmless internal betrayal, and the serious and (for the criminal network and the individual) existence-threatening external betrayal. If the choice in the pre-selection is internal betrayal, then a 'nasty' action is performed which remains invisible for the environment outside the criminal network. The consequence for the attacker (as mentioned in the last but one subsection) is to become disreputable in a similar way as it is the case with the initial normative event, provoking respective aggressive actions. The two options for external betrayal are either that the criminal provides a hint (or a criminal complaint) directly to the police, or details of the criminal network or associated activities are revealed to the public (and, hence, also to the police) by informing newspapers or other media.

Although the implementation of this part of the simulation model is quite similar in functionality to the implementation for deciding on a reaction on decreasing image (e.g. due to a norm violation, as described above), the actual implementation is quite different. The reason is that here the implementation follows much closer

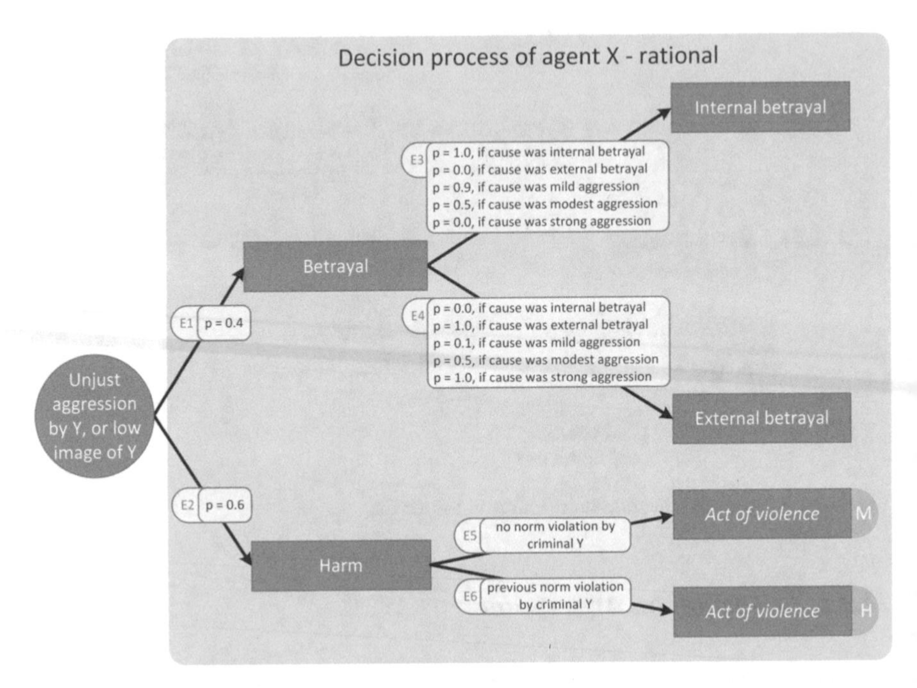

Fig. 11.15 Decision tree for rational reaction on aggression

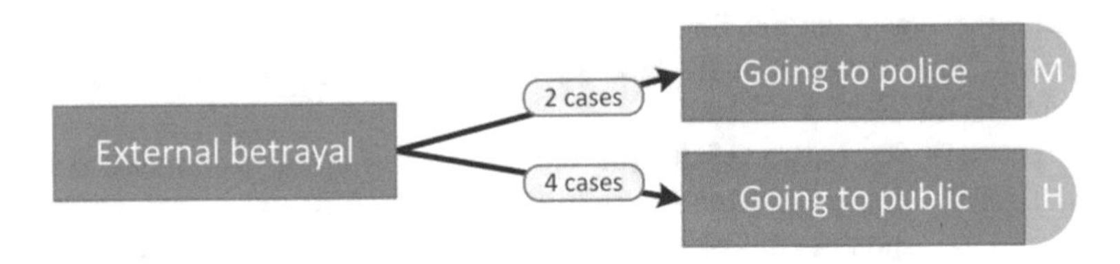

Fig. 11.16 Decision tree for external betrayal

the CCD action diagram, as more concrete evidence is available for these parts of the model.

Figure 11.15 shows the decision tree outlining the implementation for the case the criminal is in rational frame. The first stage is to decide whether to betray or to harm. Probability for the former is 0.4 (E1), for the latter 0.6 (E2).

If the decision is to betray, then the two options are selected by the conditions shown in (E3) and (E4). As a short summary, if the original aggressive action was internal betrayal, then the reaction will always be internal betrayal, too. If it was external betrayal or any other kind of strong aggression (threat or violence), then internal betrayal as a reaction is never an option. In all other cases there is a probability >0 for internal betrayal. Possible acts of internal betrayal are listed in Fig. 11.5; for external betrayal Fig. 11.16 shows the respective decision tree. The decision to inform the public results in unveiling the existence of the criminal network,

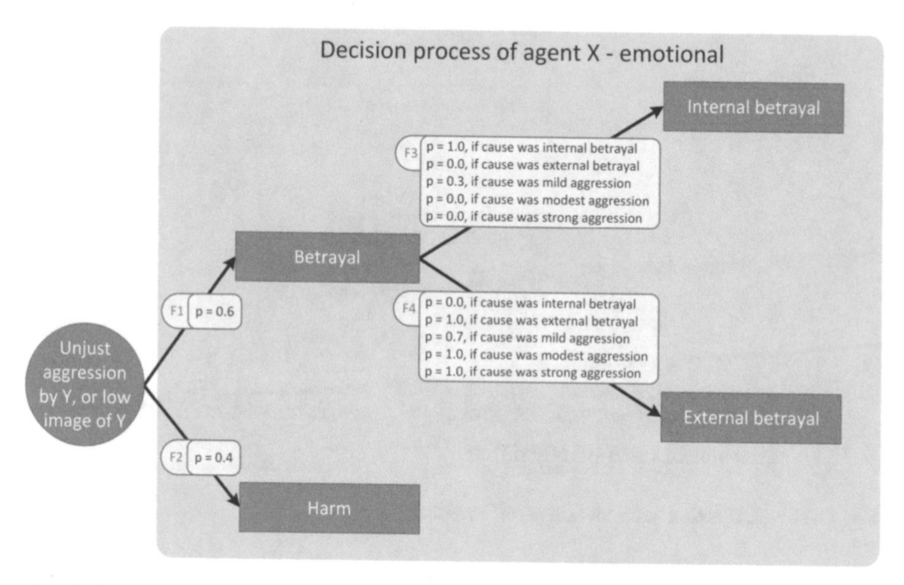

Fig. 11.17 Decision tree for emotional reaction on aggression

but also sets the grounds for having this activity regarded as a violation of the norm of trust. The consequences of informing the police are different, as this is as a covert aggression only visible to the betraying criminal and the police, who now knows about the existence of the criminal network, kicking off an investigation.

The decision to harm is just followed by the selection of an appropriate act of violence. If criminal Y is guilty a norm violation at some time in the past, then criminal X will respond with strong (highly severe) violence (E6), otherwise with modest violence (E5). The acts of violence correspond with the options in Fig. 11.7.

For the case the criminal is in emotional frame, the decision process differs slightly, as specified in Fig. 11.17. The probabilities for betrayal and harm are the other way around, i.e. 0.6 for betrayal (F1) and 0.4 for harm (F2). Also the probabilities for internal (F3) and external (F4) betrayal are different, while in the case of harm the actual violent act is selected by chance (decision tree in Fig. 11.7). In summary, the chances for more severe measures (in particular in terms of external betrayal) are higher in emotional frame than in rational frame.

11.4.7 Police Intervention

The third actor besides the Black and White Collar Criminals included in the simulation model is the police as an institutional agent. The only action modelled in the CCD action diagram is the start of an investigation because of a criminal complaint

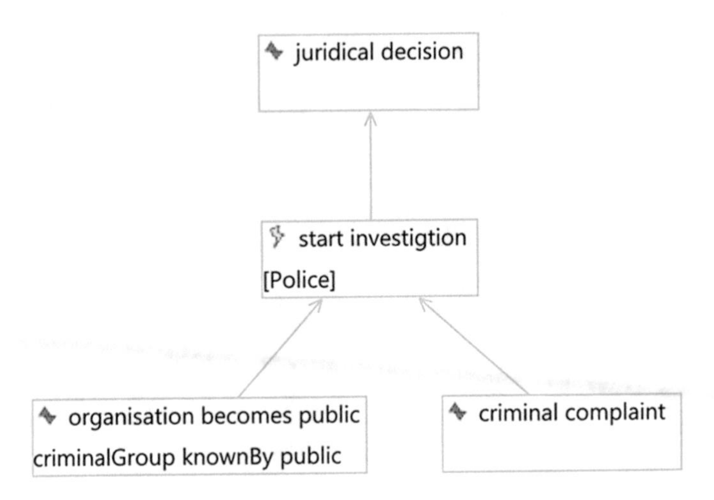

Fig. 11.18 CCD action 'start investigation' by police

or media reports, which finally results in a juridical decision, that is the arrestment of a criminal (Fig. 11.18).

The implementation of this action involves a few more aspects. Both mentioned pre-conditions trigger police activities. The first step is the generation of a report, which contains information about the reason for and the subject of investigation to initiate, as well as the source of information. Reason can be either media report or criminal complaint; subject is the criminal network. If the subject is unknown to the police so far, then a new investigation is initiated, and the investigation progress starts with 0 %. If the subject is known already, then the progress advances (with an additive calculation) by 50 %. In any case, the progress of investigation changes randomly with every tick: it (expectedly) increases by up to 40 %, but might also decrease by up to 10 %. This negative progress expresses a possible 'dead end' in which an investigation branch might enter.

If finally 100 % progress is reached, members of the criminal network are known to the police so that measures can be decided and finally taken by arresting a randomly selected criminal. An arrested criminal does no longer take part in any business of the criminal network.

11.4.8 Run on the Bank

Starting point for the effect of a 'run on the bank' is another kind of panic, the fear to lose invested money. This is modelled in the CCD as shown in Fig. 11.19. Three possible pre-conditions for this panic are envisaged: the arresting of a member of the network due to a police intervention, the knowledge about intimidating activities against the White Collar Criminal and the conjuncture that the network became public. In the implementation a slight variation is realised as the latter of these three

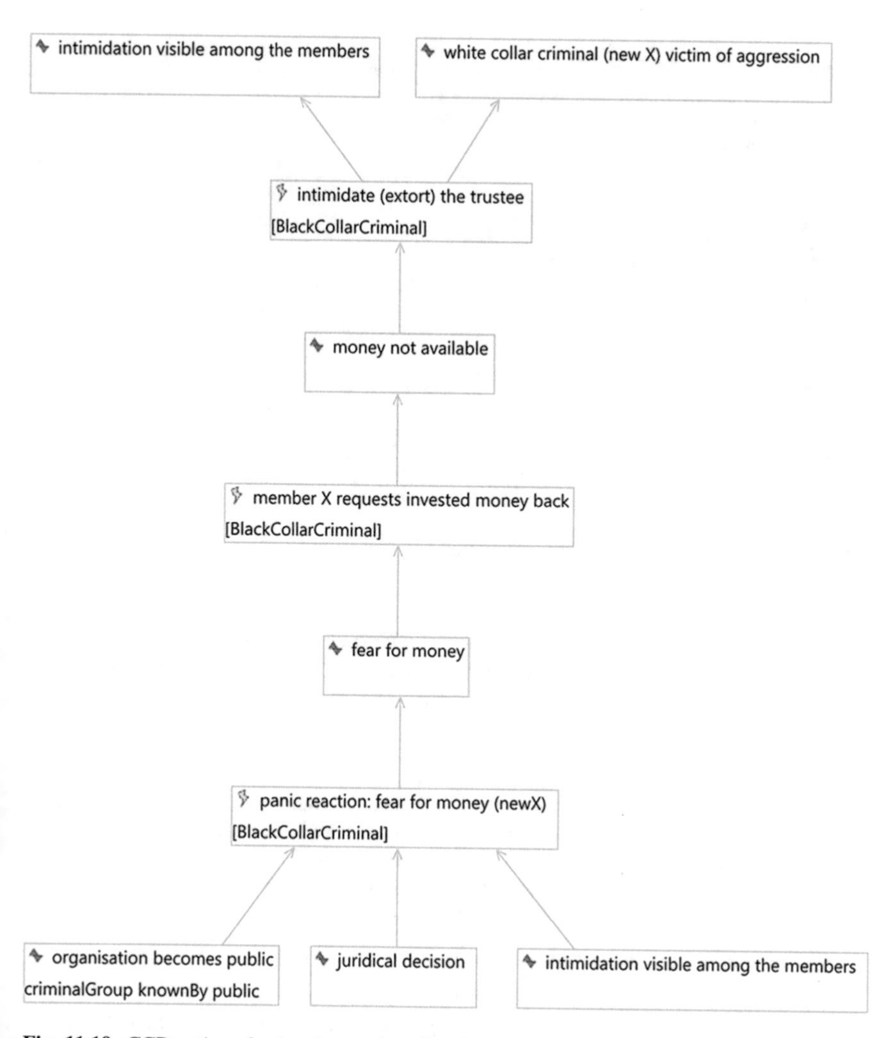

Fig. 11.19 CCD actions for 'panic reaction: fear for money (new X)' and intimidation of White Collar Criminal

conditions is not taken into account. This simplification is justifiable because the uncovered network triggers police investigations that ultimately lead to arresting of criminals, which then cause panic reactions anyway.

The panic might cause an intimidation of the White Collar criminal—in this context called the 'trustee'. This is a two-stage process, where a (Black Collar) criminal in panic to lose the invested money starts an approach to get the money back. If the White Collar is unable to return the money, the actual intimidation—often in shape of an extortion attempt—takes place. This extortion on the one hand results in aggressive actions against the trustee but on the other hand might be observed or become known by other members of the criminal network. The latter leads to an escalating number of approaches to get hold of invested capital.

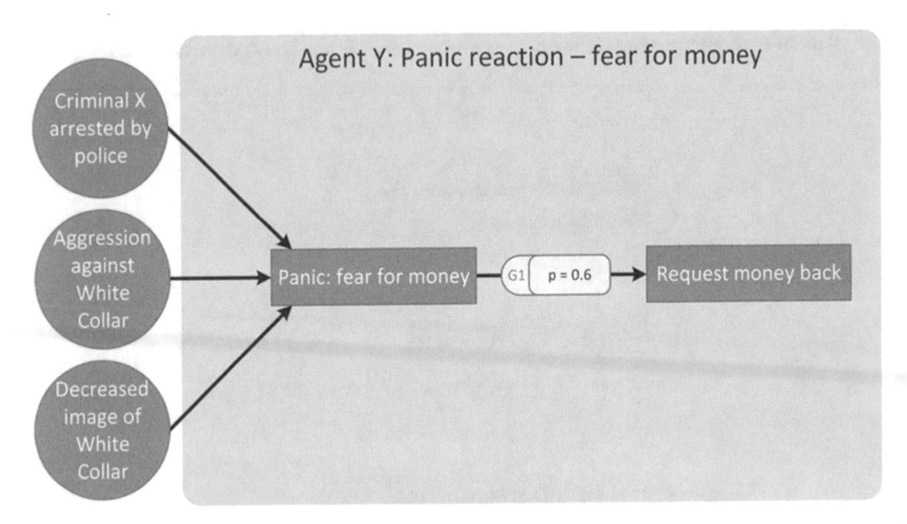

Fig. 11.20 Decision tree for panic reaction: fear for money

The entering of the panic mode is implemented as drawn by the decision tree in Fig. 11.20. The request to get the money back from the White Collar criminal as result of the panic sets in with a probability of 0.6 (G1), if one of the following conditions holds:

the state of a fellow criminal changes to 'arrested',

an aggression against the White Collar criminal is observed, or

the image of the White Collar criminal decreases to a modest or worse level. This case is not explicitly modelled in the CCD, but becomes important for the dynamics when the White Collar is involved in a conflict, and the opposite party of the conflict at some time responds with requesting the invested money back.

The mechanism with which the intimidation is implemented is deterministic. The approached White Collar criminal processes this request by trying to fulfil as many as possible of the requests appearing at the same tick. A capital stock of 20 million units is available initially, which is refilled by another 20 million units each tick after some amount was requested and paid back. The decision to pay or not to pay is communicated to the requesting criminal. For Black Collar Criminals that got their money back the crisis is resolved for the moment, and no reaction is to be expected. In the other case, the refusal to return the money is interpreted as a norm violation. Hence, a request for a normative event is issued, containing the message that the norm of trust is violated by the White Collar Criminal. This normative event triggers the mechanism of revenge or sanctioning again, as described in the beginning of the chapter.

11.5 Conclusions on the Simulation Model Description

The previous sections are intended to present the simulation model in a way to enable interested readers to comprehend the formalisations done on base of the conceptual model as well as the simulation experiments elaborated in the following sections. However, all the technical details that are inevitable for executable software systems can obviously not be presented in the frame of a book chapter. These are aspects like the configuration of parameter settings, the control of simulation runs and the generating and visualisation of simulation outcomes. In the following, a few remarks are given to each of these aspects.

Simulation parameters are implemented in three different ways:

- Parameters interesting for experimentation and typically without relation to the evidence base can be put on the Repast user interface. This is done for the number and relation of reputable and ordinary Black Collar Criminals.
- Parameters that have a close relation to the evidence base are typically modelled in the conceptual model and annotated with phrases from the evidence. Hence, these parameters have to be changed in the CCD, and a following code transformation updates the parameters in the simulation model. For example, the types and probabilities of the aggressive actions are modelled in this way.
- All other parameters are coded in the rules, in most cases as probabilities, with comments in the source code.

To run a simulation, the procedure typical for RepastJ 3.1 simulation models has to be followed. Since DRAMS is just a software framework used from within the Repast/Java code, it is basically transparent to the user.

During simulation runs outputs are generated which are presented and stored in different ways. The DRAMS rules produce text statements, written to a console window and stored in a log file. Per run there is also a sequence diagram generated and stored in an UMLet[2] file, showing all the interactions that appeared in the run. Finally, a graphical visualisation of the agents with animations of the events happening in each tick is presented while running a simulation. All these different representations are base for the results presented in the following section.

---

[2] http://www.umlet.com/.

11.6 Simulation Results

11.6.1 Narrative of the Scenarios: A Virtual Context for Real Possible Courses of Action

Simulation models typically generate an output such as times series or histograms. Here the output is different: A simulation run generates a story which describes a scenario. In the following some of these scenarios generated by model runs will be described. The objective of the scenarios is exploring the fact that the development of the behavioural rules of the agents is based on a qualitative analysis of textual data. The rules that are fired during the simulation runs can be traced back to annotations in the original textual documents. Examples of these 'open codings' have been shown in Chap. 10 describing the analysis of the textual data. In the description of the scenarios the rules are now traced back to the original annotations in order to develop a narrative of the simulation runs, i.e. the scenarios are a kind of *collage* of the empirical basis of the agent rules. Thus the reader will find text elements that had already been used to illustrate the conceptual model. However, a different composition of single pieces of evidence (generated by the execution of the program code) generates different stories. Firing of certain rules makes certain follow-up actions more likely whereas others are excluded. By exploring the behavioural space of the model the scenarios attempt to explore counterfactual situations of a complex configuration in which many decisions are involved that make different outcomes likely. This can be described as 'virtual experience'. For this reason the scenarios develop a storyline of a virtual case. (See Corbin & Strauss, 2008 for the notion of a storyline that provides a coherent picture of a case. They treat the storyline as the theoretical insight of a qualitative analysis.) In sum, the scenarios close the cycle of qualitative simulation, beginning with a qualitative analysis of the data as basis for the development of a simulation model and ending with analysing simulation results by means of an interpretative methodology in the development of a narrative of the simulation results.

For this reason, the description suggests to be a story of human actors for exploring the plausibility of the simulated scenarios. The plausibility check consists of an investigation whether the counterfactual composition of single pieces of empirical evidence remains plausible, i.e. if they tell a story. Nevertheless, the reader should be aware that the story is about software entities which execute rules programmed in the code. Italics in the text indicate that the description paraphrases annotations of the empirical text basis of the fired rules. The scenarios explore the path dependency of the simulation runs generated by probabilistic decisions rules. The only variation of the parameters is the number of so-called 'reputable' and 'ordinary' criminals (see description of the model). RC stands for reputable criminal, C for ordinary criminal and WC for white collar criminal, who is responsible for money laundering. The scenarios presented here do not represent the full behaviour space of the model but only those that are of interest for examining modes of conflict regulation and outbreak of violence in a group with properties comparable to the empirical case, namely a group with no managerial authority assigned to certain positions such as a 'boss' or 'godfather' in a professional, mafia type organisation. Nevertheless individuals differ in their reputation. We show cases that are representative for certain typical classes of the course of simulation runs. First a scenario is presented that resembles central features of the data. Second, this is contrasted by a simple example of how escalation of violence could have been avoided. The third scenario represents a case in which the group managed to overcome a severe escalation of violence. Finally the fourth scenario shows a case of successful police intervention.

Scenario: Eroding of a Criminal Group by Increasing Violence

The scenario consist of 7 RC, 3 C and 1 WC.

Tick 2–Tick 15: Initial Violence

The drama starts with an external event. For unknown reasons C0, who never was very reputable, became susceptible. It might be due to an unspecified norm violation, but it may not be so and just some bad talk behind his back. Eventually he stole drugs or they got lost. However, at least RC 1 and RC4 decided to react and agreed that C0 deserves to be severely threatened. The next day RC1 approached C0 and told him that *he will be killed* if he is not loyal to the group. C0 was really scared as he could not find a reason for this offence. He was convinced that the only way to gain reputation was to demonstrate that he is a real man. So *threw the head of RC1 against a lamp pole and kicked him* further on more when he sank down to the ground. RC1 did not know what was happening to him that such a freak as C0 was beating him down; RC1 is one of the most respectable men of the group. There could only be one reaction: He pulled his gun and shot. However, while shooting from the ground the bullet missed the body of C0.[3] So he was an easy target for CO. He had no other choice than pulling out his gun as well and shot RC 1. Figure 11.21 shows this initial sequence of escalating violence as it is displayed in the visualisation of the simulation model.

Tick 16–Tick 17: Spreading of Mistrust

However, this gunfight decisively shaped the fate of the gang. When the news circulated in the group hectic activities broke out: WC *bought a bulletproof car* and C1 thought about a *new life on the other side of the world, in Australia.* In panic RC6 wanted to severely beat off the offender. While no clear information could be

[3] Note that the following description slightly deviates from the story developed in the simulation: In the simulation the agents reason about the aggression. In contrast here, there is immediate shooting, which might be regarded as 'ad-hoc' reasoning. Similar events can be found in descriptions of other cases of fight between criminals as for instance the Sicilian Cosa Nostra (Arlacchi, 1993).

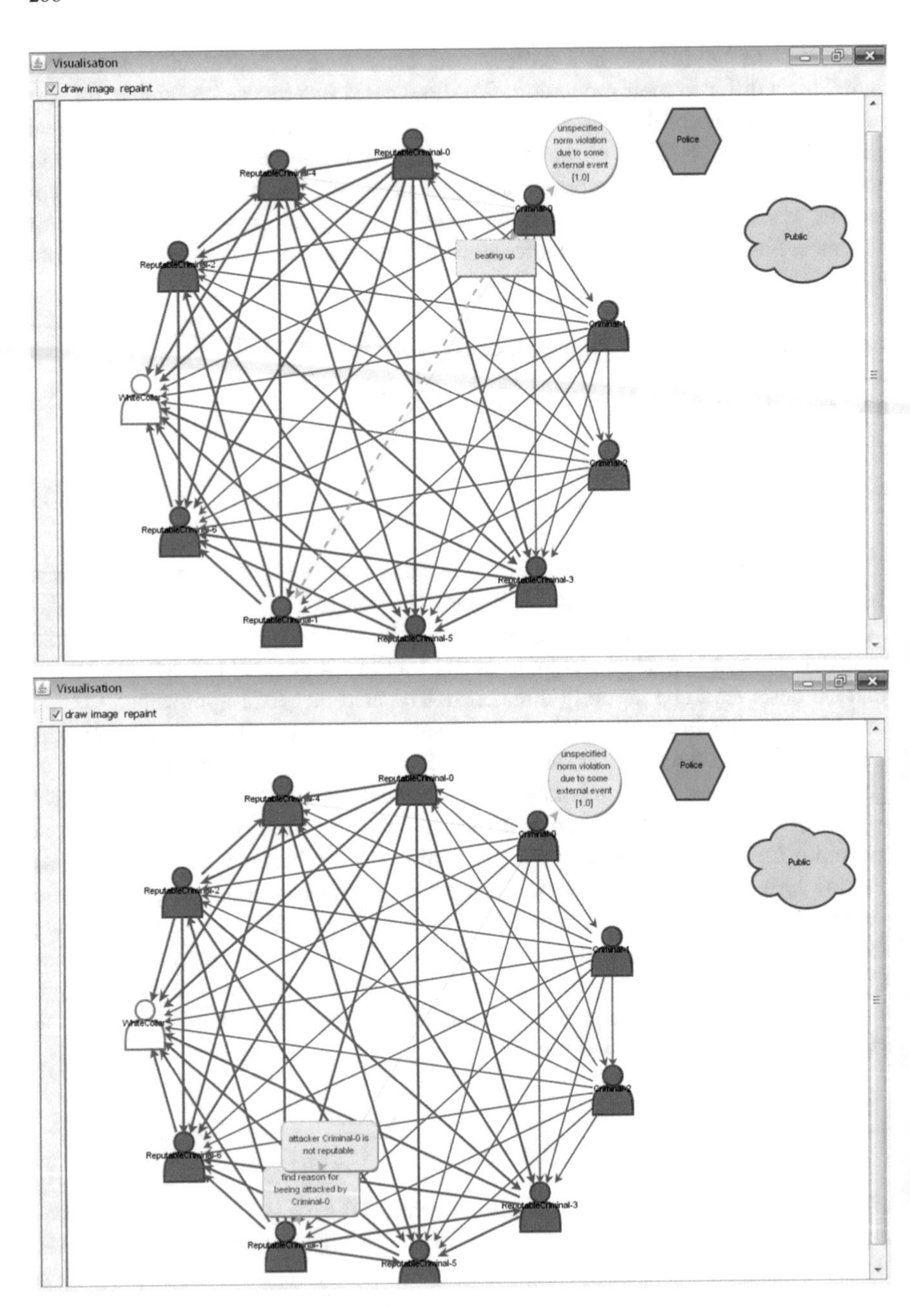

Fig. 11.21 The initial gun fight

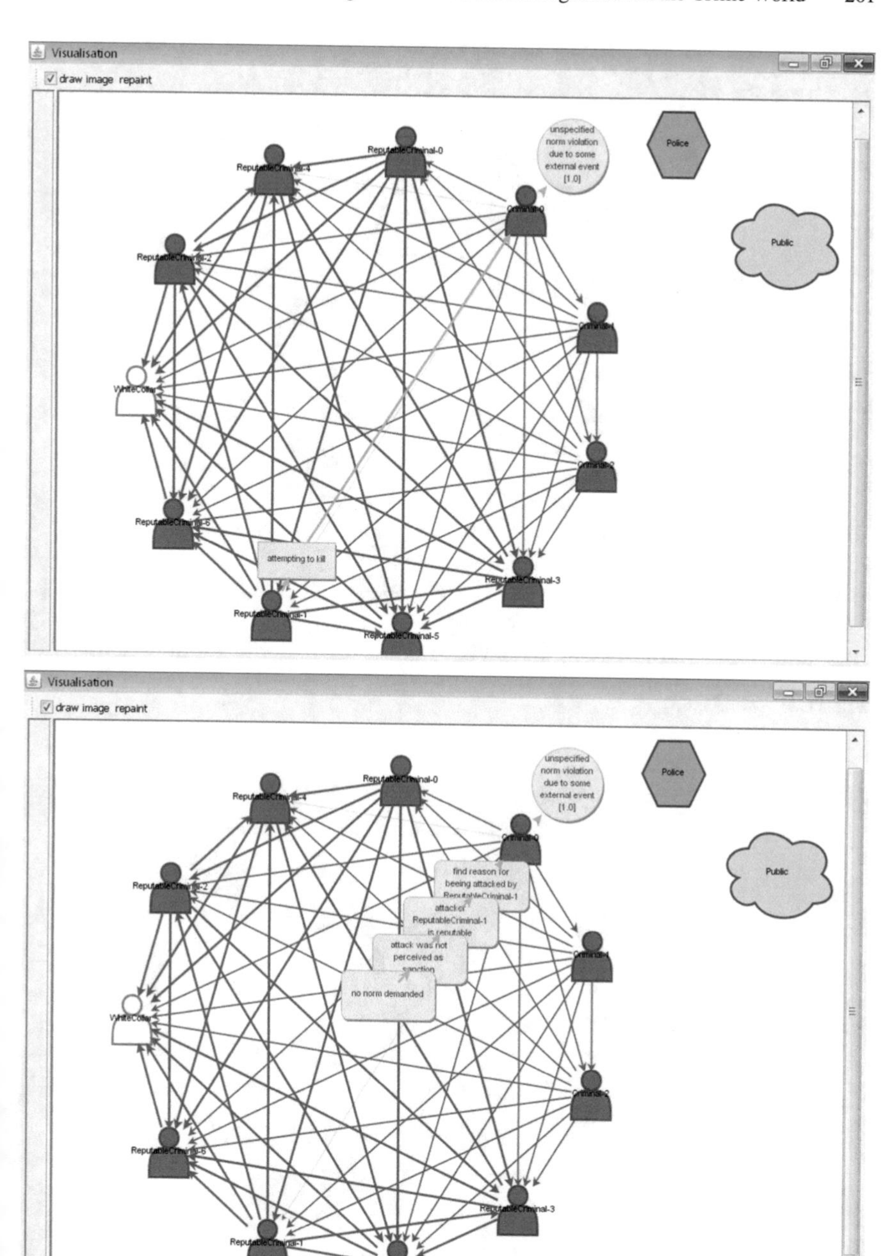

Fig. 11.21 (continued)

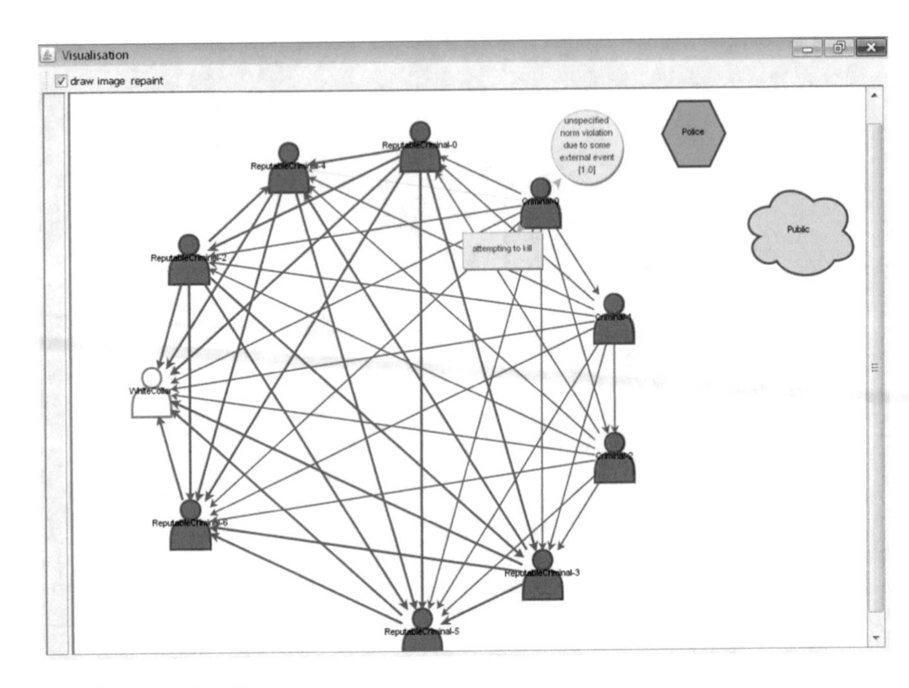

Fig. 11.21 (continued)

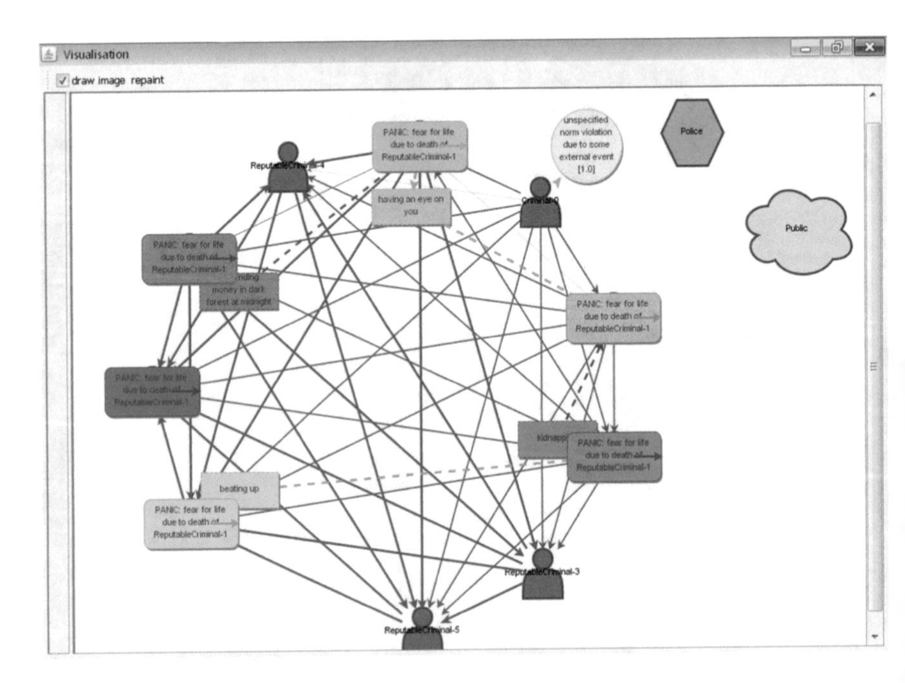

Fig. 11.22 Panic after the gunfight

obtained he presumed that C2 must have been the assassin. So with brute force he beat the hell out of C2 until he was fit for the hospital. *His head was completely deformed, his eyes blue and swollen.* At the same time, RC0 and C2 agreed (wrongly) that it was C1 who killed RC1. While C2 argued that they should kidnap him, the more rational RC0 convinces him that a more modest approach would be wiser. He went to the house of C1 and told him that *his family would have a problem* if he ever will do something similar again. However when he came back RC2 was already waiting for him: with *a gun in his hand he said that in the early morning he should come to the forest* for handing out money. Figure 11.22 is an exemplary screenshot of how this panic is displayed in the model visualisation.

Tick 18–Tick 35: Increasing Panic

But now all the victims are scared: RC0 and C1 and C2 thought about the aggression but find no norm demanded by their offenders. RC0 was fed up with being attacked by his *old friend* RC2 and invoked the general public as audience to articulate his disappointment: *in an interview with a major newspaper he betrayed his role in the network.* However, also C1 learned quickly. When he read the interview he *contacted the newspaper and told them about the role of RC0.* Meanwhile C2 planed his revenge: He *contacted a contract killer* to murder RC6, but the alleged professional turned out to be an amateur: the assassination was a failure. However,

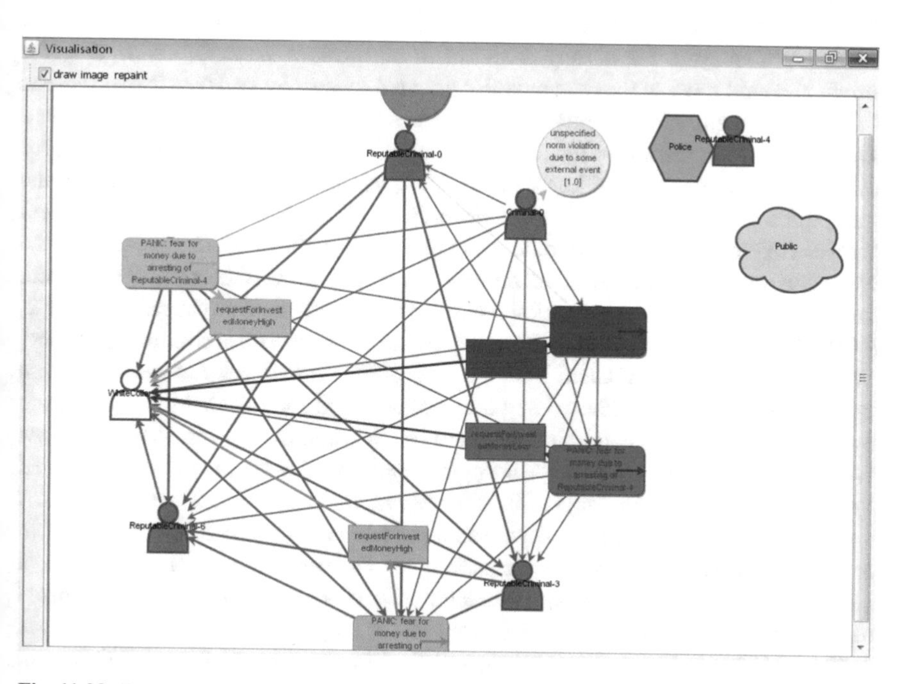

Fig. 11.23 Panic after arresting of RC4

at least RC6 could not identify C2 as purchaser of the killer and was unable to counteract. At the same time RC2 was wondering why his old friend RC0 betrayed him in the news and thought that he deserves a severe beating. At the next occasion he slammed his face. Now *remarkable tensions in the relation between RC0 and RC2 broke out.* RC2 was *really in fear.* Secretly *he wrote an anonymous letter to the police* and the police started an investigation of the case. However, it took a while until they were able to collect sufficient evidence for action. For quite a long time nothing seemed to happen. The group went back to its usual business and it seemed that peaceful relation had been restored.

Tick 36–Tick 41: Beginning of Extortion

But the silence was only an illusion. Unexpected by the criminals at one day the police arrested RC4. Figure 11.23 displays the panic after the arresting of RC4. This put the final nail in the coffin: Retrospectively one can say that in this moment *a corrupt chaos* broke out. As they realised that their secret had been disclosed all criminals were in fear for their money. In panic RC2, C1, C2 and RC5 attempted to get their money back. However, after WC paid back RC5 *he had serious problems with his liquidity* and was not able to fulfil the demands of the others. They debated what to do and decided that RC2 was best suited to enforce their claim: he went to the office of WC and *told him that he should repay his debt, because otherwise he will be killed.* However, rumours spread in the group that WC is about to be killed. This had a serious side effect. Now the others got in panic and started *extorting WC.* Indeed WC obeyed. *He arranged a deal in an offshore financial centre to get a credit* and paid most of the demands. *He paid but at least he survived.* Nevertheless, RC3 and C2 came away empty-handed and enforced their claims. *WC was now in a completely despaired situation and tried to counteract. He hired an outlaw gang for murdering RC3* because he knew that *in the gang many hated RC3.* However, the gang did a bad job and he survived. Nevertheless RC3 was shocked as he thought that his standing in the group would make him untouchable. As WC realised that the assassination failed he obeyed the demand for money. Nevertheless RC3 retaliated, but also his attempted assassination remained unsuccessful. This caused RC6, RC2, RC0, C1, and C2 to try to get their money back as long as WC is still living. Indeed, WC made a deal with C1 and RC2 by *selling them a building far below the true value.* When C2 and RC0 heard that they got outraged and decided for a plan made by RC0: He arranged an appointment between WC and his lawyer. *However, when WC entered the office, RC0 was waiting for him instead of the lawyer, accompanied by two seemingly Russian guys. One of them ordered him on his knees and pressed a machine gun at his head. … Then they forced him to sign a contract that he is no longer the owner of his investment company …. WC was in great fear for life. From that moment on, when he has appointments with RC0 he was wearing a bulletproof jacket.* And still the extortion of WC went on for longer. RC2 ordered him to *come in the night to the forest near the town and still wanted more money.* WC *was so*

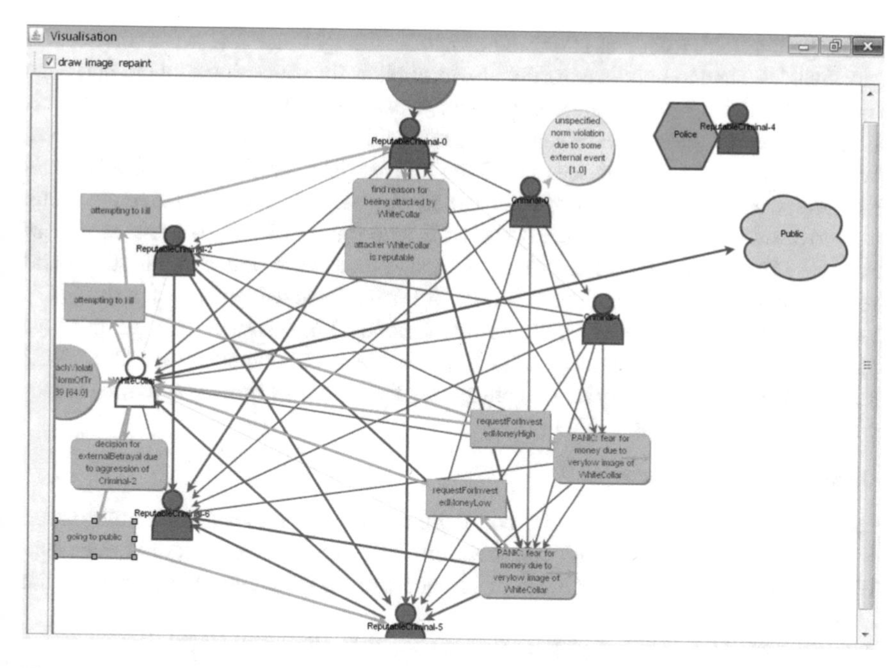

Fig. 11.24 WC going mad

much in stress that *he looked like years older.* He wondered why RC2 threatened him because usually RC2 was a reliable guy and he paid him already.[4]

Tick 41–Tick 73: Rampage of WC

It was a vicious cycle: The more he got extorted, the more urgently did all others demand their money back. It was like a run on the bank. After rumours arose of RC2 threatening WC, also RC0, C1, RC3, and RC5 made claims. WC *borrowed some money from a friend* for RC5 but refused the claims of the others. Now *he was like a hunted cat…. Isolated from the rest of the group* he *became completely hysteric.* This situation is shown in Fig. 11.24.

WC wondered if he should kidnap RC6 but then he made a different plan. *He called to meet RC2 at the construction site for a discussion. However, in fact he came with two weapons. With one weapon he wanted to shoot down RC2 and put the other weapon in the hand of RC2 to claim that he shot only in self-defense.* When RC2 saw him with the gun he *pulled out a machine gun and tried to hold it in his stomach* but WC was faster and shot him to death. Still outraged he wanted to go on and also kill C2 but this time he failed. But still his feelings of vengeance were not satisfied. He arranged *an interview in a newspaper* in which he made severe accusa-

[4]The space characters ('…') indicate that two different in vivo codes (at different parts of the original text) had been used.

tions against RC5. RC5 was shocked because he thought that WC was a trustworthy guy. Still WC had to handle extortion. Several plans for squeezing him out had been made. C2 and RC 3 now completely lost any trust in WC. While being in fear for his life, their fear for money was even stronger and they requested money back. Indeed, WC found a way to give money to RC3 but refused the claims of C2. Instead he *told the newspaper also the crimes* of C2. Now all were in panic and the fate of the group was governed by a *rule of terror*. It was RC3 who finally *killed WC in the middle of the street* in front of his office. After the assassination the police captured C0 and several haphazard plans had been made. Still many wanted revenge for the WC's rampage and tried to find a way to get their money. For instance RC3 did not give up a plan for kidnapping. But after a while the group faded away. Only RC2 remained silent. Eventually he still enjoys the fruits of his criminal activities somewhere in the South Seas.

Decisive Critical Junctures

Initial gun fight: That the bilateral conflict escalated in murder caused outbreak of panic in the overall network and therefore diffusion of the conflict in the group.
Wrong assignment of perpetration of offence caused spreading of violence in the group.
Arresting of RC4 leading to fear for money: this started the cycle of extortion whereas initially WC was not involved in the conflict.
Rampage of WC: turned the panic from fear for money to fear for life.
Killing of WC blocked restoration of the group.

Scenario: A Small Irritation

The scenario consists of 3RC, 7C, and 1 WC.

Tick 1–Tick 4: Initial Loss and Restoration of Trust

At the beginning of this story RC5 began to mistrust WC. Eventually, WC embezzled his money as RC5 *had invested a significant amount of black money in WC's company structure*. However, it remains ambiguous what exactly happened. Anyway, RC5 wanted his money back. *They had a meeting at the office of their lawyer to appraise the value* and WC obeyed the request. So trust was restored and

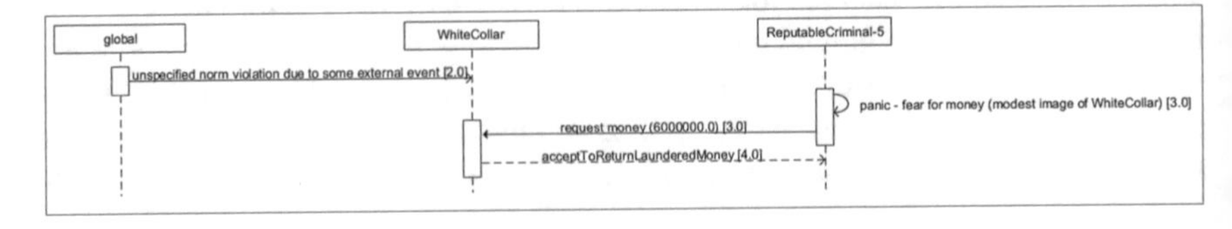

Fig. 11.25 Sequence diagram of the scenario

the groups continued their criminal activities. The sequence of actions is displayed in Fig. 11.25.

Decisive Critical Junctures

Initial loss of trust against WC (and not another criminal) provided the chance for conflict resolution (by paying) and to avoid escalation of the conflict.
WC obeyed the request for money.

Scenario: The Group Overcomes Severe Escalation of Violence

The scenario consists of 7RC, 3C, and 1 WC. In order to make the description of the scenarios more compact we abstain from presenting further graphical illustrations in the following.

Tick 2–Tick 7: Brute Force

At the beginning of the story C1 had been accused by RC1, RC2 and RC5 of having violated their trust. It remained unclear what exactly happened. However, they agreed that C1 should be under observation and RC5 *installed concealed microphones and even a camera in his apartment.* But C1 realised that they mistrusted him and *was in fear of being monitored.* He had a strong ego and even though he knew that RC5 was a respectable man he could not endure such an affront. Without hesitation he *shot RC5 to death in the middle of a busy avenue.*

Tick 8–Tick 19: Spreading Mistrust

This sudden excessive violence completely out of proportion was a shock. RC0 and C2 got in panic. RC0 *bought a bulletproof car.* C2 attempted retaliation of the murder of *his longtime ally*. He was convinced that such an exorbitant murder, much like an execution, could only be mandated by RC1, the *arch-enemy* of C1. He knew that RC1 would come to a big party at the next weekend. There he waited for him with some of his comrades. What followed was *like a mafia movie: they approached him and slammed his head when he wanted to enter the party room. In panic RC1 ran out of the building. One of the guys threw him against a street lamp but he could escape in the dunes directly behind the building. There he wanted to hide but the goons were behind him. He ran to the street and jumped in a taxi. The taxi driver brought him to a hospital.* In fact C2's suspicion was simply wrong. So RC1 had no idea what was happening to him, but he swore that C2 will be sorry for his offence. However, his revenge was more sophisticated. He knew that C2 was responsible for a major drug transport and he stole a considerable amount of the commodity. Nevertheless, as C2 realised that he was betrayed he suspected that RC1 was behind it. As he was more a goon he decided to ultimately solve the problem by shooting him to death. *He called to meet RC1 at the construction site for a discussion. However, in fact he came with two weapons. With one weapon he wanted to shoot him down and put the other weapon in his hand to claim that he shot only in self-defense.* However, RC1 anticipated that the meeting would be a trap. Instead of coming to the meeting *he gave an interview with a major newspaper in which he provided detailed insights in C2's criminal activities.* So the conflict between C2 and RC2 finally resulted in a disclosure of the secrecy of the group. In fact, the police started a criminal investigation.

Tick 20–Tick 27: WC Becomes Involved

As secrecy is obviously essential for undisturbed drug dealing, RC3 wanted to give him a lesson and *started an affair with his girlfriend.* Indeed RC1 felt cuckolded and was wondering why RC3 betrayed him. At the same time, C2 got outraged when reading the news and retaliated in kind. He contacted the *newspaper and told them about RC1's role in the group.* Immediately when he read it, RC1 went to the home of C2 and *hold a pistol against his head,* shouting '*now you will die!*' Also C0 was fed up. He wanted to kill C2 but something cropped up: The police investigation resulted in arresting R6. As it now became clear that the police was pursuing them, all were in fear of losing their investments and tried to get it back as soon as possible. Now WC was in trouble. RC0, C1, C0 and RC4 *came to his house and ordered that he should come at 10 in the wood near the town. There they asked for money.* In fact, *threatening and intimidation* worked: WC paid as much as he could. *I paid but I'm alive* as he later said. However, it took not long until he got *problems with his liquidity* and he was unable to pay RC0 and RC4.

Tick 28–Tick 37: Police Interferes in Ongoing Violence

The bank refused the monetary transfer because of the negative account balance. RC0 made a phone call to WC. He was really angry. In their favourite club C0, C1 and RC3 saw how he ran out of the café *with lather in his mouth and kicked a bike against a tree.* This made them wondering how save the rest of their money was. They took their *standard approach: threatening and then ask for the money.* However, now WC was curious. What the hell did they want furthermore? He said to them that he *does not know how to pay anymore because his bank account was completely empty.* Not much later *the police received an anonymous letter.* One may wonder who wrote the letter ….

At the same time conflict between RC1 and C2 that occupied the group for a long time already was still not resolved. RC1 *hired an outlaw motor cycling gang to assassinate* C2. However, they did not do a very professional job. C2 survived the attack and *told the newspaper* that he already had contacted previously about the attempt. Meanwhile the individual reactions to the crisis created more and more a *corrupt chaos.* While C0 started making plans to kill WC because he thought that *WC wanted to keep for himself their investment* WC wanted to take on initiative himself. *He was at a point where he was totally despaired.* He got guns and had a plan to shoot down RC0. *He made an appointment with RC0 at the construction site for a discussion. However, in fact he came with two weapons.* However, in the last moment he didn't dare because *he was afraid of fingerprints at the gun.* Yet as RC0 saw WC approaching him with a weapon he was so much in fear for his life that he did not counteract. However, as rumours spread telling this story RC1, C0, RC2 and RC3 lost trust in WC and demanded their investment back. RC1 told WC that *his family will die if he does not pay.* In fact, WC sold an apartment to C0 but *for a price that was much too low.* However, he did not serve the other demands. They were not amused. However before they could do anything C0 got arrested.

Tick 37–Tick 55: Severe Extortion of WC

In panic all business partners wanted to extort WC ever more now. RC1 arranged an appointment between WC and his lawyer. *However, when WC entered the office, RC0 was waiting for him instead of the lawyer, accompanied by two seemingly Russian guys. One of them ordered him on his knees and pressed a machine gun at his head* to enforce their claims. Furthermore on his way home RC4 laid and waited for WC. He fired a gun to shoot him down but the bullet missed the target. As this news spread, also RC2 threatened him to death and RC3 and C1 took their *standard program: threatening and then ask for the money.* RC3, RC1 and RC0 shared their job and every day when WC came home from work one of them *was already waiting for him at the front of the door of his house.* WC *was like a hunted cat* and in great *fear of his daughter*. However, he had no liquid money. He made deals with C1, RC1 and RC2: *he signed certificates that transferred the ownership of his investment company* to them and *RC1 became its new director.* In consequence the rest of the gang became even more nervous and intensified the pressure on WC. Quite some time the whole group was completely occupied with intimidating and extorting WC. For instance, he was *kidnapped three times* and held in arrest for several hours, *several times he was threatened to death*, and more than once he was beaten until *his head was completely deformed, his eyes blue and swollen.* He was lucky that he survived all the attacks. He undertook several tricks to get money such as *letting apartments owned by his company for half of the price* and *tried to get a mortgage from the offshore market.* He even *asked a friend for money* but could not fulfil all requests. Friends said that *he looked years older.*

Tick 56–60: WC Strikes Back

While WC accepted that the demands were justified finally intimidation was too much. He *secretly contacted the police. … He was shocked when he was told by the police that they knew that he was on a death list.* However, presumably his reaction was not like the police expected. Outraged he *contacted an outlaw gang and gave them the rest of the money he had to kill his enemies.* As many of them *hated some members of the gang* for a long time they undertook a massacre. On their bikes they drove to the favourite club of the gang and with heavy machine guns they fired haphazardly in the pub until the room was full of blood and impact holes. It lasted only a minute until they drove away with full speed and left back the dead bodies of RC1, RC3 and RC4. This was a shock. Never has such brute violence been observed before and all survivors wondered what actually happened. However, WC smartly erased any traces to him: He remained in *contact with the police for several times* and *gave a public interview* in which he completely laid open the criminal operations of the group. However, at the same time he agreed to a financial deal with RC0 in which he *bought fictitious rights for a major infrastructure construction.* He got a bank loan for that deal. To pretend a prestigious business *the meeting was held at a lake in Switzerland and WC came in his own private jet.* However, the *whole project was just fictitious.* Thereby WC succeeded to preserve his appearance both to the police and the criminal group. But then C2 got arrested by the police and in panic RC0 and RC2 requested their investment back. So his plan failed.

Tick 61–Tick 71: Restoring the Business

RC0 undertook another attempt to kill WC and the spiral of intimidation and extortion seemed to start again. WC was afraid that his plan will be disclosed and obeyed the request. He paid at least to C1 but could not pay RC2, who asserted that *he will kill him if he doesn't pay.* In spite of WC's partial cooperation RC0 *launched intimidating pictures of WC to the media.* However, before anybody could undertake any further action the police intervened and arrested RC0. Now worry about the money was more urgent than personal animosities. After RC2 *placed a machine gun in front of his stomach* WC paid RC2. *The deal was financed by redeeming mortgages on a construction in Curacao* and all agreed that debts have been settled. Nothing happened any more. Even though many had been killed and arrested, the police could not break up the network. Step by step the remaining members of the group restored trust and build up their business model again. Eventually they still sell drugs to the street hawkers until today.

Decisive Critical Junctures

Wrong assignment of perpetration of offence caused spreading of violence in the group.
Police intervention leads to involvement of WC in the conflict.
Double-faced counter-reaction of WC could have restored trust (if no further police interventions would have happened).
WC's final acceptance of requests enabled (possibility of) restoration of trust.

Scenario: Successful Police Operations

The scenario consists of 3 RC, 7C, and 1 WC.

Tick 2–Tick 3: Brute Force

The beginning of the story remains unknown. C4, who never was very reputable, became susceptible to C3 and C6. It might be due to an unspecified norm violation, but it may not be so and just some bad talk behind the back. It must have been a severe offence since C3 and C6 agreed that a severe reaction was in need. While C3 argued for threatening him to death, C6 was convinced that a death penalty would also be a sign to the overall group. He *hired an outlaw gang* for assassinating C4 and *they shot him to death.*

Tick 4–Tick 9: Spreading of Violence and Mistrust

However, the reaction of the group was different than C6 expected. For instance, RC1 *bought a bulletproof car* in panic. However, as it remained unclear who mandated the assassination the reaction remained ambiguous too: C1 who *was for more than 15 years a friend of* C4 presumed that RC2 was guilty and beat the hell out of him until he was fit for the hospital. *His head was completely deformed, his eyes blue and swollen.* On the other hand, C3 suspected that RC0 mandated the assassination. As revenge *he planned to approach him with a weapon but in the last moment he didn't dare.* Now C2 and RC0 were scared as they *didn't know what was happening to them. A witness testified that RC0 said that C3 must be crazy*. While for some time they remained silent, RC0 was so frightened that he *wrote an anonymous letter to the police* nevertheless. Also C2 planned revenge. On the next occasion he paid C1 back in kind: he wanted to kill him, but *the attack was betrayed* and C1 was *able to escape to Italy.*

Tick 10–Tick 17: Police Starts Intervening

For quite some time it seemed that peaceful relation had been restored and the group went back to its ordinary business. They thought *things were going well and finalised some quite successful projects.* But they didn't know that the police was after them and still C1 wanted revenge. Feeling safe abroad *he gave an interview with a major newspaper and betrayed the role of C2 in the criminal group*. Finally, the police arrested C3. As it seemed to be obvious that the arresting of C3 was the fault of C1, C0 *became completely hysterical*. However, it was RC1 who was able to scent out C1's hideout and he shot him to death. Nevertheless, as the police operation made clear to the group that their criminal activities had been detected, all were in fear for their monetary investments and attempted to get their money back as soon as possible. Indeed, WC was able to *pay several millions to RC0* but soon *he got problems with his liquidity.* The others were not satisfied and intimidated him. After some discussion what would be the best strategy: RC1 went to WC and told him that *he will be killed if he does not pay.* Yet, at the same time the news of the killing of C1 shocked the group: But since nobody knew the assassinator, the reaction was no more than a shot in the dark. C5 and C6 approached RC2 with a weapon, but in the last moment he was able to run for cover and draw his gun himself. It ended up in a gun fight that all survived.

Tick 18–Tick 38: Police Cracks the Group

Police investigations revealed that *a huge amount of black money had been transferred in the company of WC*. The police was able to collect enough evidence to arrest him. Therefore the group kept silent for quite some time. However, the gun fight still occupied the participants. C5 didn't quit his plans for killing RC2. *He paid a huge sum to an outlaw gang in order that they should kill him* but also these guys didn't succeeded. C6 on the other hand *had several talks with the police*. So nothing seemed to happen for a while but *the alliance was deeply shattered*. RC2 took revenge and *launched some compromising pictures of C5 to the media*. However, C2 noticed it. Going to the public was a severe violation of trust. Therefore C2 arranged an appointment between RC2 and his lawyer. *However, when he entered the office, C2 was waiting for him instead of the lawyer, accompanied by two seemingly Russian guys. One of them ordered him on his knees and pressed a machine gun at his head* in order that he should never do this again. However, it was impossible to restore trust in the group. However, soon after his attack C2 was arrested by the police and both C5 and RC5 were too scared by their experience of being threatened to life. Independent of each other they decided to secretly quit the group and thought that their only chance to survive would be to *secretly contact the police*. In fact, their collaboration enabled an arrest of R0 and to break the criminal activities of the group.

Decisive Critical Junctures

Misleading interpretation of sanction (death penalty) generates outbreak of chaotic violence.
Police intervention leads to involvement of WC in the conflict.
Arresting of WC terminated extortion and the business model of the group.
The fact that many contacted the police (i.e. secretly changed sides) cracked the group.

11.7 Conclusion: Central Mechanisms

The following conclusions are drawn from simulation experiments which may not be valid for and easily transferred to a real-life context. The attempt is to highlight central mechanisms that can be found throughout the scenarios. First and foremost the ambiguity of violence stimulates its spreading in the group. Only the case of an initial loss of trust (a random event) against WC provides the chance to preserve the operations of the group before violence gets out of control. Only WC has the resources to restore trust by generous repaying as compensation for the loss of trust. This enables encapsulation of initial mistrust.

Moreover, bilateral conflicts may be long-lasting without affecting the overall group. However, they become dangerous when others become involved. This need not be the case immediately. However, in this case escalation of violence easily gets out of control. Once violence spreads WC is the most vulnerable criminal. In particular police interventions rope WC in the internal conflicts. This stimulates escalation of conflicts. For this reason, police operations directly against WC are most effective by destroying the business model. If no further extortion of WC is possible internal conflicts get reduced. Police operations are most successful if a significant number of group members change sides and cooperate with the police.

References

Arlacchi, P. (1993). *Mafia von innen—Das Leben des Don Antonino Calderone*. Frankfurt a.M: S. Fischer Verlag.

Corbin, J., & Strauss, A. (2008). *Basics of qualitative research* (3rd ed.). Thousand Oaks: Sage.

Lotzmann, U., & Meyer, R. (2011). *A declarative rule-based environment for agent modelling systems*. In The Seventh Conference of the European Social Simulation Association, ESSA 2011, Montpellier, France.

Lotzmann, U., Neumann, M., & Möhring, M. (2015). *From text to agents—Process of developing evidence-based simulation models*. In 29th European Conference on Modelling and Simulation, ECMS 2015.

Lotzmann, U., & Wimmer, M. A. (2013). Evidence traces for multiagent declarative rule-based policy simulation. In *Proceedings of the 17th IEEE/ACM International Symposium on Distributed Simulation and Real Time Applications (DS-RT 2013)*, IEEE Computer Society. pp. 115–122.

North, M. J., Collier, N. T., & Vos, J. R. (2006). Experiences creating three implementations of the repast agent modeling toolkit. *ACM Transactions on Modeling and Computer Simulation, 16*(1), 1–25.

Sabater-Mir, J., Paolucci, M., & Conte, R. (2006). Repage: REPutation and ImAGE Among Limited Autonomous Partners. *Journal of artificial societies and social simulation, 9*(2).

Part V
Synthesis and Conclusion

Chapter 12
Calibration and Validation

Klaus G. Troitzsch, Luis G. Nardin, Giulia Andrighetto, Áron Székely, Valentina Punzo, Rosaria Conte†, and Corinna Elsenbroich

12.1 Data Scarcity, Calibration and Validation

The aim of this chapter is to summarise the problems incurred during the phases of calibrating and validating the extortion racket models used by the GLODERS project. The chapter starts with the discussion of the data availability and summarises shortly the contents of Sect. 4.3. It continues with a discussion of what parameterisation, calibration, sensitivity analysis and validation have to do with each other and ends up with a discussion of the validity of the GLODERS models.

---

† Author was deceased at time of publication.

K.G. Troitzsch (✉)
Computer Science Department, Universität Koblenz-Landau,
Universitätsstr 1, Koblenz, Rheinland-Pfalz 56070, Germany
e-mail: kgt@uni-koblenz.de

L.G. Nardin, Ph.D.
Institute of Cognitive Sciences and Technologies (ISTC), Italian National Research Council (CNR), Via Palestro, 32, Rome 00185, Italy

Schuman Centre for Advanced Studies, European University Institute, Fiesole, Italy
e-mail: gnardin@gmail.com

G. Andrighetto, Ph.D.
Institute of Cognitive Sciences and Technologies, Italian National Research Council (CNR), Rome, Italy

Schuman Centre for Advanced Studies, European University Institute, Fiesole, Italy
e-mail: giulia.andrighetto@istc.cnr.it

Á. Székely, Ph.D.
Institute of Cognitive Sciences and Technologies, Italian National Research Council (CNR), Via San Martino della Battaglia 44, Rome 00185, Italy
e-mail: aron.szekely@istc.cnr.it

C. Elsenbroich et al. (eds.), *Social Dimensions of Organised Crime*, Computational Social Sciences, DOI 10.1007/978-3-319-45169-5_12

12.1.1 Data Scarcity

As discussed in Sect. 6.3, empirical data about the behaviour of extorters, victims, police and the public are rather scarce, even for countries and regions where extortion is a frequent phenomenon, as is the case particularly in Southern Italy. In Italy extortion rackets are a special category in criminal statistics. The annual report of the Italian National Institute of Statistics for 2010 (published 17 January 2012 (Istituto Nazionale di Statistica, 2012)) reports a total of 5992 extortions all over Italy of which 3271 or 57.3 % had a known offender. The first number reflects, of course, only those extortions which came to police notice and cannot include undetected extortions. Given that in 42.7 % of the detected extortions the offender was not known one can only conclude that victims often report extortions to the police without giving the extorter a name (although they usually know their names). The crime clearance rate of 57.3 % for extortions is high as compared to, for instance, arson (888 out of 9622 or 8.8 %) but low as compared to money laundering (75.0 % of 1344) or unintentional homicide (76.3 % of 1765) or receiving stolen goods (82.9 % of 23,686). And if one takes the dark figure of the mafia part of Italy's gross domestic product—16 % according to Pinotti (2012a, 2012b)—as an estimate of the dark figure for extortions, their total increases to about 7000 per year in Italy, and the clearance rate decreases to about 47 %.

Another source of vagueness lies in the propensity of interviewed persons to refuse answers to questions related with their propensity to report a crime to the police when they were involved in it, particularly when not reporting is a crime in itself—which is the case with corruption (worldwide) and with extortion (in Italy). Unfortunately (as already discussed in Sect. 9.2) extortion does not seem to have ever been the subject of surveys, but one can take the experience from corruption in Eurobarometer 79.1 (European Commission & Brussels, 2013) as a proxy for the response behaviour of interviewees who would have been asked for their readiness to report extortion. The Italian subsample consisted of 1020 interviewees of which 919 said that they had no experience with corruption (and 8 said that they did not know—whatever this means). 93 persons had either refused to answer or witnessed or assisted in an episode of corruption ("assistito" in the sense of "attended" or "participated" in the Italian questionnaire, the English codebook has "experienced", but as "vissuto", "seen", is the alternative what is meant here is that a yes to this question means that the interviewee was actively participating in the corruption, either accepting or requesting a bribe). And out of these 93 as many as 27 (or 29 %) spontaneously refused to answer the question whether they had observed or participated

V. Punzo
University of Palermo, Palermo, Italy
e-mail: valentinapunzo@libero.it

C. Elsenbroich, Ph.D.
Department of Sociology, Centre for Research in Social Simulation, University of Surrey, Guildford, Surrey GU2 7XH, UK
e-mail: c.elsenbroich@surrey.ac.uk

in an episode of corruption, and out of the 66 who admitted that they had observed or participated in an episode of corruption six refused to answer whether they had reported this event and ten said that they had reported. Hence we have 50 unreported episodes of corruption out of 66 documented episodes plus 33 supposable and probably also unreported episodes. Although the absolute numbers of corruption episodes in Italy are low, the frequency of unreported episodes and of refused answers is so high that the probability of estimating the clearance rate or the percentage of unreported cases is not reasonable for the crime of corruption, and it would not be reasonable for the crime of extortion either if the respective data were available.

Comparable statistics for other European countries are not much better. In Germany, for instance, extortion racketeering is not even a category in the annual crime statistics published by the Federal Criminal Police Office (Bundeskriminalamt, 2014a, 2014b) as there is either a category of "extortion" at large or a category of "robbery/extortion resembling robbery" of different kinds of businesses such that this statistic cannot be compared to the Italian statistic. Anyway, the German 2013 report mentions two groups mainly involved in extortion rackets in Germany: outlaw motorcycle gangs and mafia-like organisations both of the Italian and the Russian type.

Outlaw motorcycle gangs (Hells Angels, Bandidos, Gremium and Mongols) were reported with 11 cases of extortion (Bundeskriminalamt, 2013); their overall activities (including other kinds of crime) make up for less than 10 % of organised crime in Germany.

Italian mafia groups were reported with 11 cases, where in most of these cases extortion was not even reported.

Russian-language organised crime groups were reported for 30 cases, making up for less than 5 % of organised crime in Germany (Bundeskriminalamt, 2013). The most recent report (Bundeskriminalamt, 2014a, 2014b) mentions only outlaw motorcycle gangs in the context of violent extortions, but the total number of this kind of crime investigations was still low (with 23).

Thus it is not a surprise that for European Union (EU) regions other than Southern Italy reliable statistics are missing, and—as discussed above—even for this part of the EU where such statistics could be meaningful, they are of restricted reliability.

12.1.2 Parameterisation, Sensitivity Analysis, Calibration and Validation

The simulation models described in this book were more or less generated from anecdotal knowledge about extortion racket systems typical for Southern Italy. As such they originated as mental models that were formalised into computer simulation models. If one considers these computer simulation models as models of a theory in the sense of the "non-statement view" of contemporary philosophy of science (also known as the "structuralist program" (Balzer, Ulises, & Sneed, 1987)) the computer simulation model of a theory of extortion racket systems (as of any theory, see for instance (Troitzsch, 1994)) can be used as a proxy for the state-of-the-art definition of a model of this theory, containing all terms necessary to formulate the theory, namely sets of objects, functions yielding measurable features of objects and relations defined over Cartesian products of sets of objects and formulas expressing conditions that have to be satisfied ("laws"). Sets of objects are defined as agent types in agent-based modelling, functions yielding measurable features are the instance variables defined for these agent types, relations are defined as additional object types or as additional instance variables and laws are defined as programme invariants. For any theory, according to the "non-statement view", a set of intended applications needs to be defined which in the case of the theory of extortion racket systems is the set of observable extortion racket systems in different parts of the world. It goes without saying that not all terms of the theory are observable or measurable before the theory has been provisionally accepted—as the discussion of undetected cases showed there might be some terms which only become measurable with the help of the theory in a way that a certain intended application—a province in Southern Italy—satisfies the theory only under the condition that the percentage of undetected cases has a certain value or lies within a certain range of values. From what was already seen in Sect. 8.4 the province of Trapani at the western coast of Sicily is a candidate for an intended applications of the theory behind the event-oriented NetLogo model, as is the neighbouring province of Palermo for the GLODERS-S model.

This anticipated, we can give the terms in the headline of this section a more precise meaning:

Parameterisation is the act of assigning values to what one usually calls global variables in a computer programme when these are used as parameters in functions.

Sensitivity analysis is the process of finding out which of these parameters matter, i.e. which parameters influence the output metrics of a simulation run. Sensitivity analysis is only necessary and reasonable in stochastic models, as in deterministic models every change in every parameter will lead to differing output metrics.

Calibration is then the selection of parameter combinations which lead to output metrics which match the respective metrics of intended applications.

Validation is the task of finding out whether at least one or few parameter combinations lead to output metrics with the available intended applications—whose set will usually be finite whereas the set of possible parameter combinations might be infinite (in theory, not in practice as only a finite number of simulation runs can be produced and as the set of "real" numbers in a computer is also finite).

The theory of extortion racket systems described in this book is a stochastic theory which uses stochastic functions to describe the relations between agents. Hence the output metrics mentioned in the above definition are distributions rather than distinct values. This makes it necessary to discuss what the meaning of a match between the output metrics of a simulation run (a distribution function defined over the space spanned by some instance variables of the agents) and the empirical statistic of a certain regional extortion racket system is. The theory, formalised in the simulation model, states that given its parameterisation a certain output metric is distributed according to certain distribution, and the question to be answered for the sake of validation is whether the feature measured for a certain regional extortion racket system lies within an appropriate confidence interval of this distribution.

An additional problem arises from computer simulations in so far as these—other than analytical mathematical derivations—do not yield a closed formula of a distribution function but only a list of values drawn from a hopefully large sample of simulation runs which in turn allows for an approximate formula for a probability density function or for a cumulative density function such that a statistical test becomes possible.

12.1.3 Validity of the Norm-Oriented ERS Simulation Models

The validity of the GLODERS-S model has already been discussed by Nardin et al. (2016) with the result that this version yielded a precise replication of the empirical data of the province of Palermo. For the model discussed in Sect. 8.4 with its high number of 1280 runs a more detailed analysis of the validity is possible. Section 8.4 already showed a diagram with representations of all runs and of the empirical findings of the seven provinces covered by the Sicily and Calabria database (Frazzica et al., 2015). To be able to analyse the distribution of the simulation runs in more depth, the data were normalised with the help of a linear transformation in order to have the mean of both transformed variables at 0.0 and to have their variances as 1.0. Moreover the runs were weighted with the natural logarithms of the numbers of extortions that were counted per run. These numbers ranged from one to more than a thousand, and these numbers are distributed according to a lognormal distribution with mean about 80; the input parameters explain slightly more than one-half of their variance and the influence of most individual input parameters on the number of extortions is highly significant on a 0.005 level. Nevertheless this output metric was not used for the analysis proper, as its empirical counterpart is very dubious due to the high dark figure.

The resulting approximate probability density function (an exponential function of a polynomial of the two transformed variables up to fourth order[1]—a normal distribution would have had a PDF as an exponential function of a polynomial of the two transformed variables up to second order) can be seen in Fig. 12.1.

[1]The algorithm used for the approximate calculation of the probability density function was first described by Cobb (1978), extended by Herlitzius (1990) to multivariate distributions and, for instance, used by Troitzsch (1998). The algorithm works only for standardised variables and yields the best possible approximation between the empirical moments up to fourth order and the same moments of a probability distribution described with a density function which is an exponential function of a polynomial in two variables up to fourth order. Hence the diagram in Fig. 1 does not show the "true" distribution but a quite good approximation as the comparison between colour shades, contours and scattergram shows. Contours are for values of the probability density function which are multiples of 0.015; its maximum is 0.307, and the value of the cumulated density function within the outermost contour line is 0.95 which means that Trapani, Reggio ID="ITerm239"Calabria and Palermo lie within the 95 % confidence region.

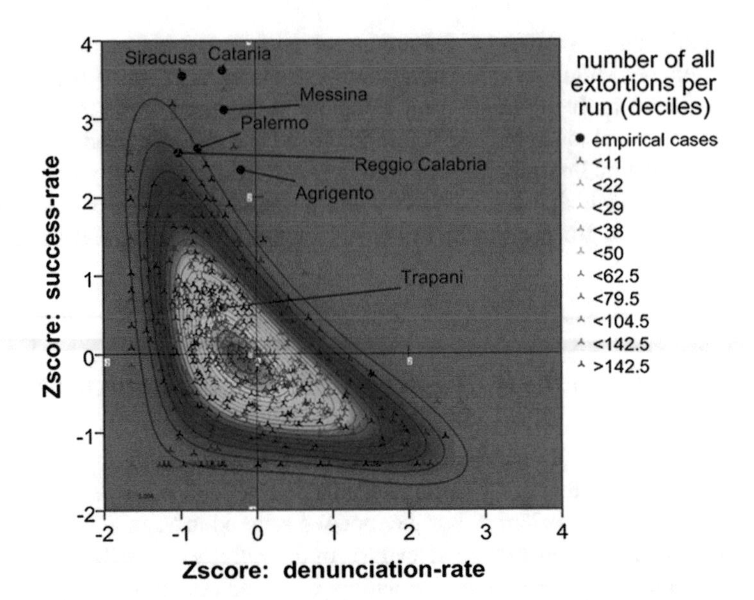

Fig. 12.1 Distribution of a Z transformation of the two main output metrics of the event-oriented NetLogo ERS model (contour curves for multiples of 0.015, coloured marks for the individual runs, *black dots* for the empirical data of the seven provinces covered by the Sicily and Calabria database)

This probability density function is far from normal. The province of Trapani is represented at a position where the PDF is approximately 0.25, the provinces of Reggio Calabria and Palermo are represented at a position where the PDF is approximately 0.030 and 0.015, respectively, whereas the other four provinces lie outside the 0.015 contour curve, but—with the exception of Catania—still near the positions of simulated runs.

Figure 12.1 shows the distribution of the two output metrics for all parameter combinations generated by the pseudorandom number generator within the intervals given in Sect. 8.4, Table 8.3, but it also gives a hint at which parameter combinations should be analysed in more detail. If it is the case that the provinces of Palermo and of Trapani are the first candidates for intended applications of the theory behind the event-oriented NetLogo model, the parameter combinations of the runs whose output metrics turned out to be most similar to the empirical data should be used to find out what the distribution of the output metrics is like for exactly these two input parameter combinations. This is what is shown in Fig. 12.2. The runs with the parameterisation of the run from Sect. 8.4, Figs. 8.1 and 8.2, matching the province of Trapani are marked in green and form a distribution with a fairly small variance whereas the runs with the parameterisation of the run matching the province of Palermo are marked in red and form a distribution with a much larger variance. The main difference between the two parameterisations can be found in Table 12.1; the other input parameter values of the two original runs resembling these two provinces did not differ much; hence the means between these runs were taken (for a detailed explanation of these parameters see Sect. 8.4, Table 8.3).

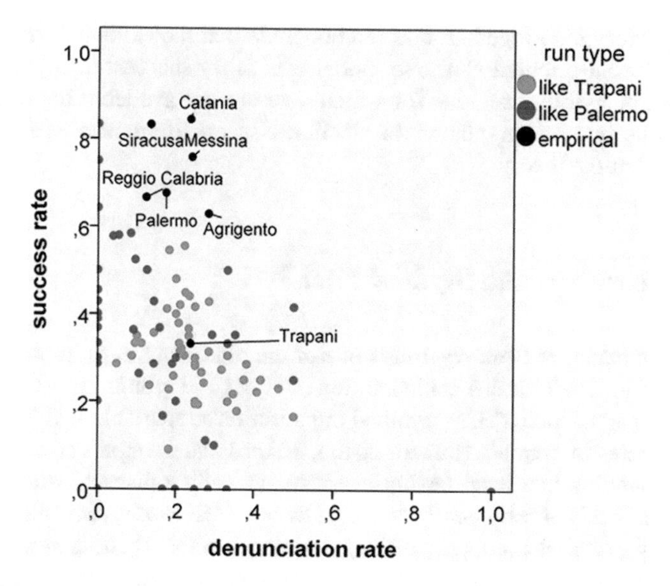

Fig. 12.2 Runs resembling empirical data

Table 12.1 Difference in the parameterisation for Reggio Calabria- and Trapani-like simulation runs

Input parameters of the two runs next to provinces	Reggio Calabria-like	Trapani-like
Cultural background	−7.25	−13.2
Communication range	2.95	5.94
Forgetfulness	0.907	0.831
Weight of the normative drive; the weight of the individual drive is 1–NDW	0.5227	0.4925
Benefit for victims	119	80
Conviction probability	.561	.331
Vision range of police	37	51
Punishment severity	22	28
Probability that a shop publishes its readiness not to denounce an extortion attempt	.061	0.074
Escape chance	0.270	0.144
Extortion radius extension	1.056	1.043

Parameterising the model with these values means calibrating it at least for these two provinces. At the same time the parameter values in Table 12.1 and the ones mentioned above are theory-supported measurements of these 11 “T-theoretical terms” (as the “non-statement view” calls these terms (Balzer et al., 1987)) for these two provinces. With even more simulation runs measuring these nine terms for the other five provinces should also be possible (the provinces of Palermo, Messina and Agrigento already found their matching simulation runs—see the red dot near the black dot for this province, better visible in Sect. 8.2, Fig. 8.2).

Putting everything together, one can conclude that the validation of the theory behind the event-oriented NetLogo model was fairly successful. It goes without saying that many other regions for which data are not available for reasons discussed in the first paragraphs of this section also qualify as successful intended applications of the theory.

12.2 Narratives and Stylised Facts[2]

The Sicilian mafia, or *Cosa Nostra*, is one of the oldest and most important Italian mafia (Santino, 1995). It is a confederation of about 150 groups[3], mostly located in the western part of Sicily.[4] This criminal organisation adopts a hierarchical structure with strict rules of conduct (Punzo, 2013a, 2013b) and is organised in cells, each one corresponding to a local territorial entity, typically a district, called *family* or *cosca*. At the basis of each *family*, you find the *men of honour*, also called *soldiers*. Each family exercises monopoly of illegal violence within their territory, checking the almost overall criminal activities (e.g. extortion, usury, drug trafficking, infiltration in public procurement) and affecting the legal economic ones (Catanzaro, 1988; Gambetta, 1993; La Spina, 2005; Sciarrone, 2009).

The Sicilian mafia is deeply rooted in society and it tends to reproduce itself over time (Gambetta, 1993; La Spina, 2008; Neumann, Frazzica, & Punzo, 2016; Sciarrone, 2009). It also provides actual protection services (Gambetta, 1993; La Spina, 2008), usually by replacing the state in the resolution of different problems. In return, the mafia increases its authority and reputation, achieving consensus on the part of society that can be even spent to mobilise the entire community against law enforcement (Punzo, 2016).

The embeddedness of the Sicilian mafia dates back to the decade between the 1950s and 1960s, when the relationship between the criminal organisation and the local political authorities strengthened. In this period, the Palermo families began to acquire increasingly more power and to infiltrate into urban development programmes.

During the 1970s and 1980s a family from the neighbourhood of Corleone and headed by the bosses Salvatore Riina and Bernardo Provenzano took power and stressed a top-down pyramidal structure, characterised by a "structure rigidly hierarchical and organised according to specific administrative models whose top positions are reached through complex selective recruitment policies [...]"

[2] Part of this chapter was published in NardinID="ITerm240" et al. (2016) "Reprinted with permission of the Journal of Autonomous Agents and Multi-Agent Systems.

[3] There is no exact number of members affiliated to the Sicilian mafia, but according to official reports, the number should be considerably lower than the estimated 3000 of the mid-1990s, perhaps as low as 2000ID="ITerm250" (see Paoli, 2014).

[4] The provinces of Palermo (the capital of ID="ITerm251"Sicily), Trapani and AgrigentoID="ITerm252" (Paoli, 2014) are particularly influenced by the Sicilian mafia.

(Di Cagno & Natoli, 2004, p. 19). Still during this period, the mafia bosses attempted to reconstruct the Territorial Commission to discuss strategic decisions and settle disputes, the so-called *Cupola* (Di Cagno & Natoli, 2004; Punzo, 2013a, 2013b; Scaglione, 2011). In contrast, especially in the 1980s and the 1990s, the state launched some important innovations in the field of criminal law, starting with the Rognoni-La Torre law.

Until the first half of the eighties, the mafia families were primarily committed to drug trafficking. The protection racket was mostly directed to those businesses that could provide significant economic revenue. Moreover, the cruelty of the Sicilian mafia in those years was expressed not only in the struggle within the organisation,[5] but also in the fight against the institutions and all the potential dangers to the organisation. In the period following the late 1980s, due to the conclusion of the Maxi Trial (Alfonso, 2011) that resulted in life sentences for almost all the heads of the leading families and an exponential increase in the number of collaborators of justice, the Sicilian mafia started a terror strategy meant to "force the institutional partners to negotiate a way out for the men of honours" (Di Cagno & Natoli, 2004, p. 43).

Directly after the massacres of Capaci and Via d'Amelio, in 1992, the fight against the mafia has gained more attention, which endured over time. The declared trend is toward increasing efforts, creating a *system* of anti-mafia measures and instruments (La Spina, 2014; La Spina & Militello, 2016), both directly by police operation or indirectly through the support to mafia's victims (La Spina, 2014; La Spina & Militello, 2016; Militello, 2013). In contrast, the criminal organisation underwent significant changes in order to identify alternative ways of penetrating the local environment (Punzo, 2013a, 2013b). The strategy of the Sicilian mafia in the management of extortions changed in favour of a systematic request, often of only small amounts, towards all the entrepreneurs, on the basis of the rule of the "*pay little, but pay all*" (Amadore & Uccello, 2009; Di Gennaro & La Spina, 2010; La Spina, 2008; Scaglione, 2008). In addition, they began a renewed attempt to expand into other areas, until then considered difficult to enter, such as to get back the leading role in drug trafficking and money laundering, and to enter the public and private construction and leisure industries and the renewable energy market. Other changes emerged with the transition of the leadership to Bernardo Provenzano, who instituted a *strategy of submersion* that led to the end of striking criminal manifestations (Di Cagno & Natoli, 2004; Dino, 2011; Palazzolo & Prestipino, 2007). This strategy has helped to reduce the strong social alarm created around the Sicilian mafia after the early 1990s, allowing the criminal group to restore its relationships with public administrators, politicians, businesspersons and professionals (Amadore, 2007; Dino, 2006).

The capture of Bernardo Provenzano in 2006 created a vacuum at the top of the criminal organisation. Since then, many convictions of its major leaders, except for Matteo Messina Denaro, who is still fugitive, have had decisive consequences. The absence of charismatic leaders, able to hold the decisional power, seems to now be

[5] During the early 1980s, the second mafia war took place. It was driven by the *family* from Corleone against other *families* from Palermo.

clear. Moreover, the new bosses show a much lower criminal capabilities than their predecessors (Punzo, 2013b).

The Sicilian mafia is currently undergoing a deep crisis and the image of a fragmented organisation is emerging. Nevertheless, it is not defeated. In fact, the Sicilian mafia remains one of the most dangerous and fearsome criminal organisations over the Italian territory and abroad. Despite the strong reaction of the state over the last 30 years against the mafia (see La Spina, 2014), the Sicilian mafia is still heavily affecting the development of Sicily (La Spina, Avitabile, Frazzica, Punzo, & Scaglione, 2013; Pinotti, 2012a, 2012b; Scaglione, 2015). This might be imputed to the fact that while legislation has been highly effective at imprisoning mafia members (see law n. 356, 1992), seizing their properties (see: law n. 109, 1996; law n. 296, 2006; law n.92, 2008; law n.40, 2010) and creating the conditions for the emergence and thriving of intermediary organisations, such as Addio Pizzo[6] and Libera[7] (law n. 44, 1999), it has been much less successful at changing citizens' behaviour and motivating them to collaborate with the state. This may be the consequence of the enduring existence of the Sicilian cultural context and some social norms that lead people to accept the presence of mafia and not to report its activity. Such social norms would then undermine the efforts of the state, creating a context in which laws fail to motivate citizens.

Here we analyse five historical periods from 1980s to 2000s of the Sicilian mafia broadly characterised by certain strategies of the different actors involved in the dynamics of the phenomenon as described above using the Palermo Scenario simulation model presented in Sect. 5.1. In particular, we describe a simulation experiment aimed at understanding the dynamics of protection rackets and explore the consequences that various policies, specifically a legal approach, typically pursued by states, and a social norm approach, have on the mafia and citizens in this complex social system.

In Sect. 12.2.1, we describe the experiment we performed to explore the five historical periods of the Sicilian mafia. Next, we discuss the results of some simulations used for analysing the historical perspective of this organisation in Sect. 12.2.2.

12.2.1 Experiment

In this experiment, we explore and analyse five historical periods of the Sicilian mafia described below. Each historical period can be broadly characterised by certain strategies of the different agents, for instance before about 1992 the mafia employed a violent strategy while following this period it undertook a more hidden strategy. Agents' strategies are linked to the simulation model via specific input parameters. In Table 12.2 we display the agents, the main strategies that they use

[6] http://www.addiopizo.org.

[7] http://www.libera.it.

and the possible values of the strategies that we consider here. In Table 12.3 we display the parameters of the model through which the strategies are implemented along with their description.

We characterise the evolution of the Sicilian mafia and the relevant agents (see Chap. 5) in the following way:

P1: This period represents the situation before 1980, in which the state had few specific legal mechanisms to fight the mafia (weak legal norms and no social norms). The mafia demanded a high amount of pizzo from entrepreneurs. Those who did not acquiesce had a high probability of being strongly punished (violent

Table 12.2 Description of the possible strategies of the agents

Agent	Strategy	Value
State	Legal norms	*Weak legal norms*: the state lacks legal mechanisms for countering the mafia or only does so weakly
		Strong legal norms: the state has legal mechanisms and uses them effectively against the mafia
State and ntermediary Organisation	Social norms	*No social norms*: agent cannot spread lawful social
		Social norms: agents can spread lawful social norms
Mafia	Racketeering	*Violent racketeering*: characterised by the high amount of pizzo -protection money-requested and strong and likely punishment for refusal
		Hidden racketeering: characterised by low amount of pizzo requests from a greater proportion of entrepreneurs and rare and mild punishment for refusal

Table 12.3 Input parameters of the strategies

Strategy	Input parameter	Description
Legal norms	`numPoliceOfficers`	Number of police officers
	`captureProb`	Probability of capturing a mafioso if the police observes a pizzo request or punishment
	`convictionProb`	Probability of convicting a mafioso after capture
	`percTransferFondo`	Percentage of mafia's resources allocated into victim-support fund
Social norms	`propCitizens`	Proportion of the population who receive a message invoking the NEW set of norms
Racketeering	`extortLevel`	The proportion of entrepreneurs' endowment requested as pizzo
	`punishSeverity`	The amount of punishment inflicted by mafiosi on entrepreneurs
	`punishProb`	Probability of punishing a non-payer entrepreneur

mafia). Additionally, much of the Palermitan population, the entrepreneurs and consumers, still held a traditional view on the mafia, in which pizzo is broadly perceived as a *legitimate* payment for protection services (Gambetta, 1993; Varese, 2013).

P2: This period represents the situation between 1980 and 1992, in which the state instituted several new coercive anti-racket laws in order to counter the mafia. These new anti-racket laws allowed the police and judiciary to more effectively counter the mafia and also allowed the state to provide support to mafia victims (strong legal norms). However, the state did not promote lawful behaviour among the population through non-legal means, such as invoking social norms.

P3: In the early 1990s, however, due to the effectiveness of the state strategy (strong legal norms), the mafia changed its violent and combative strategy (violent mafia) into a more moderate strategy in order to operate hidden from the law enforcement (hidden mafia). Concretely, this strategy stipulated a reduced amount of pizzo demanded of entrepreneurs but from a larger number of them. In addition, it inflicted a lower punishment on those that did not pay.

P4: In the mid-1990s, these changes, especially the state policies, paved the way for the emergence of civil society organisations (represented as the intermediary organisations) responsible for promoting lawful behaviour by aligning social norms with the legal norms among the population.

P5: After 2000, the state realised that to counter the mafia the use of legal mechanisms alone is not enough and began to act in order to explicitly promote lawful behaviour through non-legal means (social norms). This is reflected in the support and encouragement of initiatives that transmit these values in schools and among the general public (e.g. Festival della Legalità).

The input parameter values that we use to define the policies and characterise the periods, along with the agents and their strategies, shown in Tables 12.2 and 12.3, are informed by qualitative data extracted from empirical data analysis conducted by the University of Palermo in Sicily during the GLODERS[8] project (La Spina, 2014; Militello, La Spina, Frazzica, Punzo, & Scaglione, 2014). These data were collected through interviews with shopkeepers who had paid or been requested to pay pizzo, and analyses of judicial trials and confiscated mafia documents.

Additionally, we assume that entrepreneurs and consumers have five specific social norms. Entrepreneurs have social norms (N1) Pay pizzo request, (N2) Do not pay pizzo request, (N3) Report pizzo request and (N4) Do not report pizzo request. Consumers have the social norm (N5) Avoid paying pizzo Entrepreneurs. Norms N1 and N4 (TRADITIONAL set of norms) are part of the set of norms that are associated with the traditional mentality of the individuals regarding the mafia, in which pizzo should be paid and not reported to the police (omertà). Conversely, norms N2 and N3 (NEW set of norms) represent the set of norms that correspond to a recent emerging anti-racket sentiment that is based on the

[8] http://www.gloders.eu/.

understanding of the social and economic harm caused by the mafia. Differently to these, norm N5 is one factor that is used by Consumers to rank the different Entrepreneurs that may buy a product from.

The simulation experiment was conducted using the simulator GLODERS-S[9] developed under the project GLODERS. The experiment was run with 200 consumers, 100 entrepreneurs, 1 state, 1 mafia and 20 mafiosi. The number of police officers varies depending on the strategy adopted by the state (weak or strong legal norms, see Table 12.4).[10]

The simulation corresponds to a continuous run of the model for 50,000 time units with configurations exogenously changing at runtime every 10,000 time units. These configuration changes correspond to the sequence of periods from P1 to P5 and cumulatively take into account the results of the former periods carried over to the next.

We repeated the simulation ten times and the results were analysed based on the arithmetic mean value of the behavioural output metrics shown in Table 12.5 and the saliences of the output metrics concerning the norms.

12.2.2 Analysis

The results, displayed in Fig. 12.3, demonstrate that the introduction of anti-racket laws—between period P1 (weak legal norms state) and P2 (strong legal norms state)—drastically reduces the total number of pizzo requests ($p = 1.082 \times 10^{-05}$)[11] (Fig. 12.3a). This reduction in payment can be attributed to an increase in the state's efficiency leading to a greater proportion of imprisonment ($p = 1.082 \times 10^{-05}$) (Fig. 12.3d) and consequently larger number of imprisoned mafiosi. Nonetheless, the proportion of paid pizzo increases ($p = 1.299 \times 10^{-04}$) (Fig. 12.3b), meaning that among those (fewer) entrepreneurs who are approached a greater proportion decide to pay. This can be attributed to the increase in the salience of the set of TRADITIONAL norms that occurs during P1 and that remains stable during P2 (see Fig. 12.4). The strong legal approach undertaken during P2 is completely ineffective at reducing the salience of TRADITIONAL norms. Hence, even though, in period P2, the state becomes effective capturing and convicting mafiosi, it is not successful in making the NEW set of norms more salient than the TRADITIONAL ones in the entrepreneurs' mind (Fig. 12.4). This can be observed in Table 12.6 and Table 12.7.

[9] Simulator available for download at https://github.com/gnardin/GLODERSs/.

[10] The numbers of ID="ITerm253"agents of each type are arbitrary. However, we assume that five entrepreneurs per mafioso is a reasonable number to be handled by an individual. Moreover, the number of police officers ranges from 5 to 20, meaning that in an extreme case there is the same number of police officers as mafiosi.

[11] All statistical significance tests shown in this chapter are performed using the Wilcoxon rank sum test with $\alpha = 0.05$ (Hollander & Wolfe, 1973, pp. 68–75). We chose this test due to the fact that our data cannot be assumed normally distributed under the Shapiro-Wilk test (Shapiro & Wilk, 1965).

Table 12.4 Agent's strategies and parameter values according to each historical period

Period	Agent and configuration			
P1: Pre-1980	**State**		**Mafia**	**Intermediary Organisation**
	Weak legal norms	*No social norms*	*Violent*	*No social norms*
	numPoliceOfficers=5 captureProb=0.2 convictionProb=0.1 percTransferFondo=0.0	propCitizens=0	extortLevel=0.1 punishSeverity=0.75 punishProb=0.9	propCitizens=0.0
P2: 1980–1992	**State**		**Mafia**	**Intermediary Organisation**
	Strong legal norms	*No social norms*	*Violent*	*No social norms*
	numPoliceOfficers=20 captureProb=0.8 convictionProb=0.6 percTransferFondo=0.5	propCitizens=0	extortLevel=0.1 punishSeverity=0.75 punishProb=0.9	propCitizens=0.0
P3: 1992–1995	**State**		**Mafia**	**Intermediary Organisation**
	Strong legal norms	*No social norms*	*Hidden*	*No social norms*
	numPoliceOfficers=20 captureProb=0.8 convictionProb=0.6 percTransferFondo=0.5	propCitizens=0	extortLevel=0.03 punishSeverity=0.5 punishProb=0.5	propCitizens=0.0

P4: 1995–2000	**State**		**Mafia**	**Intermediary Organisation**
	Strong legal norms	*No social norms*	*Hidden*	*Social norms*
	numPoliceOfficers=20 captureProb=0.8 convictionProb=0.6 percTransferFondo=0.5	propCitizens=0	extortLevel=0.03 punishSeverity=0.5 punishProb=0.5	propCitizens=0.1
P5: Post-2000	**State**		**Mafia**	**Intermediary Organisation**
	Strong legal norms	*Social norms*	*Hidden*	*Social norms*
	numPoliceOfficers=20 captureProb=0.8 convictionProb=0.6 percTransferFondo=0.5	propCitizens=0.05	extortLevel=0.03 punishSeverity=0.5 punishProb=0.5	propCitizens=0.1

Table 12.5 Behavioural output metrics

Metric	Description
Number of pizzo requests	Total number of pizzo requests made
Proportion of pizzo paid	Proportion of pizzo requests paid by entrepreneurs
Proportion of reports	Proportion of non-paid pizzo requests that are reported to the state
Proportion of imprisonments	Proportion of mafiosi incarcerated

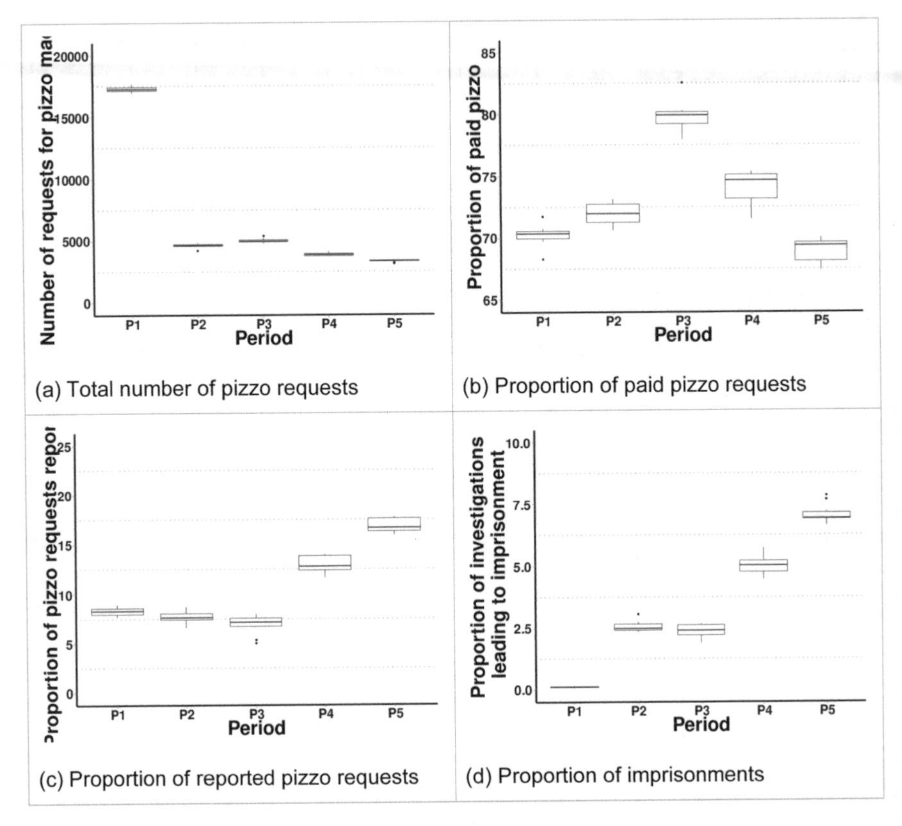

Fig. 12.3 Results of the simulation of the periods shown in Table 12.4 according to the output metrics shown in Table 12.5. (**a**) Total number of pizzo requests; (**b**) proportion of paid pizzo requests; (**c**) proportion of reported pizzo requests; (**d**) proportion of imprisonments

In period P3, the mafia responds to the state' strong approach and changes its strategy from violent to hidden characterised by requesting lower amounts of pizzo and inflicting softer punishments. As a result, the mafia successfully increases the proportion of entrepreneurs that pay pizzo ($p=1.082\times10^{-05}$) (Fig. 12.3b); a reduction in the threat of punishment and in its severity actually allows the mafia to obtain

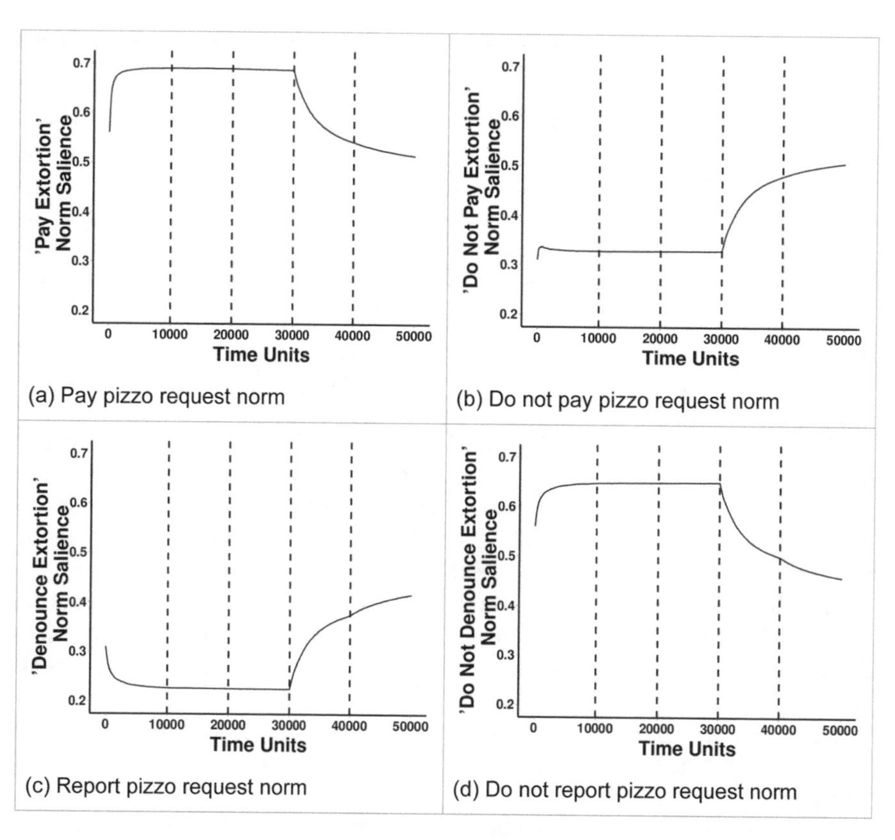

Fig. 12.4 Mean value of the entrepreneur norm salience. The *y*-axis shows the mean strength of the norms and the *x*-axis represents the elapsed simulation time measured in time units. The *dashed lines* indicate the moment in which the periods' configuration is changed beginning with period P1's to P5's configuration. (**a**) Pay pizzo request norm. (**b**) Do not pay pizzo request norm. (**c**) Report pizzo request norm. (**d**) Do not report pizzo request norm

Table 12.6 Entrepreneurs' norm salience mean and standard deviation value at the end of each period

	Periods				
Norm	P1	P2	P3	P4	P5
N1	0.689±0.02	0.689±0.01	0.687±0.02	0.542±0.02	0.514±0.02
N2	0.327±0.02	0.327±0.01	0.328±0.01	0.479±0.02	0.506±0.01
N3	0.227±0.01	0.226±0.01	0.225±0.01	0.374±0.02	0.416±0.04
N4	0.647±0.01	0.648±0.01	0.649±0.01	0.499±0.02	0.485±0.04

a greater proportion of payments. The success of the hidden mafia strategy may be partially imputed to the fact that the TRADITIONAL set of norms is still highly salient among the entrepreneurs indicating the inadequacy of the state actions in favouring a greater change on the entrepreneurs' mindset towards the NEW set of

Table 12.7 Proportion of entrepreneurs associated with each set of norms at the end of the simulation

Period	Traditional (%)	Do not pay only (%)	Report only (%)	New (%)
P1	100.0	0.0	0.0	0.0
P2	100.0	0.0	0.0	0.0
P3	100.0	0.0	0.0	0.0
P4	97.0	0.3	0.0	0.0
P5	56.4	25.6	3.4	14.6

norms.[12] This impact is presented in Table 12.7, in which the salience of the TRADITIONAL and NEW set of norms remains relatively unchanged in period P3 compared to P2.

The inclusion of the intermediary organisation in period P4, whose main activity is to promote the NEW set of norms through the social norm-based approach, however, changes the situation. As a result, we observe a reasonable decrease in the proportion of paid pizzo ($p = 1.082 \times 10^{-05}$) (Fig. 12.3b) and an increase in both the proportion of reported pizzo (Fig. 12.3c) and imprisonments (Fig. 12.3d).

Additionally, a group of entrepreneurs (about 0.3 %) shifted their dominant norm regarding the payment of pizzo from a situation in which the salience of *pay pizzo request* norm is higher to one in which the salience of *do not pay pizzo request* norm is higher. Thus the promotion of lawful behaviour performed by the intermediary organisation leads to some change in the entrepreneurs' normative mindset, which is also reflected in both the increase in reported pizzo (Fig. 12.3c) and mafiosi imprisonment (Fig. 12.3d).

Finally, in period P5, the state begins an activity that complements the action of the intermediary organisation. It starts promoting lawful behaviour among the population by encouraging the adoption of the NEW set of norms and giving more visibility to its actions and results obtained in countering the mafia. Looking at Figs. 12.3 and 12.4, we note a significant change in the proportion of paid pizzo requests with respect to period P4 ($p = 1.082 \times 10^{-05}$) (Fig. 12.3b). Analysing the transitions shown in Table 12.7, we note that complementing the action of the intermediary organisation in spreading the NEW set of norms, the state improves significantly the transition of the entrepreneurs' mindset from the TRADITIONAL to the NEW set of norms (about 14.6 %). Another 25.6 % of entrepreneurs make a partial transition and have *do not pay pizzo request* as their dominant norm, and an additional 3.4 % shifted to the *report pizzo request* norm.

This suggests that the social norm-based approach, such as the promotion of lawful behaviour, is complementary to a legal norm-based approach. The analyses of the way in which these two policies complement each other have been made possible by the use of agents endowed with complex architectures. These type of architectures allowed us to inspect agents' minds and understand how norms affected their decisions instead of relying only on the analysis of their behaviours.

[12] We are aware that several other factors may have influenced this change; however, here we model only the normative aspect.

12.3 Participatory Modelling for Validation

Sections 12.1 and 12.2 discuss availability and use of data in the GLODERS models. In addition to hard statistical data and use of narrative data or stylised facts, GLODERS employed a third data-gathering method: participatory modelling.

Participatory modelling is a form of modelling used in descriptive modelling processes such as system dynamics and increasingly in agent-based modelling (Badham, 2015). Participatory modelling uses relevant practitioners, or stakeholders, to extract their expertise of the field they work in.

Participatory modelling can use different methods to elicit this knowledge, such as interviews with individuals, focus groups with a range of stakeholders or conferences and workshops consisting of talks and discussions. The data collection can be structured, by collaboratively constructing concept or cognitive maps (Papageorgiou & Salmeron, 2013; Prell et al., 2007). However, often the data is collected in a more narrative form and the mapping later done by modelling experts.

One of the main reasons for using participatory modelling is the complexity of the target system models are built to represent (Byrne, 2013). Most often models are there to help the understanding of interdependent subsystems. Practitioners often bring high levels of expertise of the subsystem they work in. The discussion with other stakeholders can elicit how the subsystems hang together and how they interact. A second aspect that is important for model building is an idea of the temporal dynamics of the target system. Practitioners often have deep knowledge of the system over time and can provide important information about processes and causal mechanisms.

Extortion racket systems are just such a complex system, consisting of different agents and agencies: legal, normative and spatial aspects; communication sources and channels; behaviour influences; and expectations and collaborations. Extortion racketeering has previously been analysed isolating certain aspects such as the legal framework (Militello, 2011) or the perpetrator-victim relationship (Smith & Varese, 2001). And while deterrence and legal aspects are essential for an analysis of extortion racketeering, extortion racketeering is a crime that is deeply embedded in social, economic and spatial aspects (cf. Chap. 3).

GLODERS employed participatory approaches throughout the project through a stakeholder board, consisting of 27 domain experts from 10 countries, including judges, police commissioners and policy researchers in the field of extortion racketeering. Participation of the stakeholder board was written into the project at several stages and for a variety of purposes.

1. Provision of data: Extortion racketeering, as much of criminology, is beset with the difficulty of obtaining data. Stakeholders working in the field can function as gatekeepers to data otherwise unavailable. In GLODERS data access through stakeholders was integral to the model building. Obtaining detailed documentation of legal proceedings, interviews, witness statements and police investigations of extortion racketeering allowed for data-driven modelling of processes inside the criminal world (cf. Chaps. 10 and 11) and model calibration in the

society focussed model (Chaps. 7 & 8) model. Data exchange relied on mutual trust as well as the focus on models being useful for practitioners once built (see point 3).

2. Harnessing expertise: Although access to data and documentation was essential for the success of GLODERS, what was probably even more important was harnessing the knowledge of international extortion racketeering experts throughout the project. In an initial international workshop the project was introduced to the stakeholders. This was very important as computational modelling is a relatively new methodology, which most practitioners never heard of before. Introducing them to the methodology and its application in GLODERS allowed a discussion about mutual expectations and needs. For the stakeholders it was important to understand what might be needed to make GLODERS a success and what they could expect as a project outcome. For the consortium it was important to understand what stakeholders wanted and needed as results and what they were able and willing to contribute.

   Two following stakeholder meetings, after 12 and 24 months, respectively, presented intermittent modelling results to the stakeholder board to obtain feedback. Feedback received was on the identification of different relevant actors, the basic behaviours of these actors, the interplay of normative, legal and social aspects and the initial dynamics results of the models. These interim validations helped to keep the model development realistic and relevant.
3. Ensuring relevance: Relevance was indirectly ensured through the interim validation as mentioned above. Keeping a close eye on a realistic implementation made sure that the simulation focussed on understanding the target system, resisting shortcuts and simplistic operationalisations. The stakeholder meetings also had direct influence on the relevance as stakeholders questioned the applicability of the models directly. Two aspects are important in modelling, that the model is accurate and that the model is built with a purpose in mind. A model built for training police officers is very different from a model used for crime prediction.
4. Increase impact: Involving stakeholders in the modelling throughout the process does not only improve the model itself but also the dissemination of the model. The involvement ensures that stakeholders get the model they need and want, meaning they will be more likely to promote the model with colleagues and collaborating organisations. Transfer of academic research into the policy realm is not without hurdles. Involving practitioners will break down at least some of the hurdles although it will not be able to solve all (e.g. different time frames). Particularly the model of processes inside the criminal world was built in close collaboration and permanent contact with stakeholders and will be used in a practice setting to analyse extortion racketeering networks by providing a kind of virtual experience for the police. GLODERS also had a more formal way to engage stakeholders in the impact and dissemination through running a stakeholder training workshop in which practitioners were introduced to the final models with a focus on the practical aspects of model use and application.

Participatory modelling has many useful features for modelling research and GLODERS is a good case study for explicating the four listed above. Engagement with practitioners was written into the project in the form of the stakeholder board. Four formal stakeholder meetings took place, one kick-off meeting in which expertise and expectations were shared, two interim validation meetings and one training workshop. In addition to the formal interactions the consortium had regular contact with the relevant stakeholders throughout the project. In addition to the formal stakeholder board other practitioners were engaged, such as interviews with *Addio Pizzo* and *other civic organisations.* The interviews with civic organisations helped to understand how bottom-up resistance might be modelled and how it might affect both state actions (supporting law enforcement) and change in society (change of social norms).

The GLODERS models show how modelling can integrate a range of data sources, such as surveys, stylised facts, legal documents, narrative data and, last but not least, participant modelling.

References

Alfonso, G. (2011). *Il maxiprocesso venticinque anni dopo. Memoriale del presidente*. Rome: Bonanno.

Amadore, N. (2007). *La zona grigia. I professionisti al servizio della mafia*. Palermo: La Zisa.

Amadore, N., & Uccello, S. (2009). *L'isola civile. Le aziende Siciliane contro la mafia*. Torino: Einaudi.

Badham, J. (2015). Functionality, accuracy, and feasibility: Talking with modelers. *Journal on Policy and Complex Systems, 1*(2), 60–87.

Balzer, W., Ulises, M. C., & Sneed, J. D. (1987). *An architectonic for science. The structuralist program*. Dordrecht: Reidel.

Bundeskriminalamt. (2013). *Organisierte Kriminalität.*Wiesbaden: Bundeslagebild 2013.

Bundeskriminalamt. (2014a). *Organised crime. National Situation Report 2014*. Wiesbaden.

Bundeskriminalamt. (2014b). *Police crime statistics—Federal Republic of Germany—Report 2014.* Wiesbaden.

Byrne, D. (2013). Evaluating complex social interventions in a complex world. *Evaluation, 19*(3), 217–228.

Catanzaro, R. (1988). *Il delitto come impresa. Storia sociale della mafia*. Padova: Liviana.

Cobb, L. (1978). Stochastic catastrophe models and multimodal distributions. *Behavioral Science, 23*, 360–374.

Di Cagno, G., & Natoli, G. (2004). *Cosa Nostra ieri, oggi, domain*. Bari: Dedalo.

Di Gennaro, G., & La Spina, A. (2010). *I costi dell'illegalità. Camorra ed estorsioni in Campania*. Bologna: Il Mulino.

Dino, A. (2006). *La violenza tollerata. Mafia, poteri, disobbedienza*. Milano: Mimesis.

Dino, A. (2011). *Gli ultimi padrini*. Bari: La Terza.

European Commission, Brussels. (2013). *Eurobarometer 79.1. ZA5687 Data file Version 3.0.0*. Abgerufen am 26. 2 2016 von GESIS Data Archive, Cologne: doi:10.4232/1.12448.

Frazzica, G., Punzo, V., La Spina, A., Militello, V., Scaglione, A., & Troitzsch, K. G. (2015). *Sicily and Calabria Extortion Database*. (GESIS, Hrsg.) Retrieved May 17, 2016, from Datorium: http://dx.doi.org/10.7802/1116.

Gambetta, D. (1993). *The Sicilian mafia: The business of private protection*. Cambridge, MA: Harvard University.

Herlitzius, L. (1990). Schätzung nicht-normaler Wahrscheinlichkeitsdichtefunktionen. In J. Gladitz & K. G. Troitzsch (Eds.), *Computer aided sociological research* (pp. 379–396). Berlin: Akademie Verlag.

Hollander, M., & Wolfe, D. A. (1973). *Nonparametric statistical methods*. New York, NY: Wiley.

Istituto Nazionale di Statistica. (2012, January 17). *Crimes reported by the Police Forces to the judicial authority*. Retrieved March 17, 2016, from http://www.istat.it/en/archive/50621.

La Spina, A. (2005). *Mafia, legalità debole e sviluppo del mezzogiorno*. Bologna: Il Mulino.

La Spina, A. (2008). *I costi dell'illegalità. Mafia ed estorsioni in Sicilia*. Bologna: Mulino.

La Spina, A. (2014). The fight against the Italian mafia. In L. Paoli (Ed.), *The Oxford handbook of organized crime*. Oxford: Oxford University Press.

La Spina, A., Avitabile, A., Frazzica, G., Punzo, V., & Scaglione, A. (2013). *Mafia sotto pressione*. Milano: FrancoAngeli.

La Spina, A., & Militello, V. (2016). *Dinamiche dell'estorsione e risposte di contrasto tra diritto e società*. Torino: Giappichelli.

Militello, V. (2011) Crimini internazionali e principi del diritto penale, *Ars interpretandi. Annuario di ermeneutica giuridica*, 182–201

Militello, V. (2013). Transnational organized crime and European Union: Aspects and problem. In Böll-Stiftung, H. & Schönenberg, R. (Eds.) *Transnational organized crime. Analyses of a Global challenge to the democracy, Transcript* (pp. 255–266)

Militello, V., La Spina, A., Frazzica, G., Punzo, V., & Scaglione, A. (2014). *Quali-quantitative summary of data on extortion rackets in Sicily*. Deliverable 1.1, GLODERS Project.

Nardin, L. G., Andrighetto, G., Conte, R., Székely, Á., Anzola, D., Elsenbroich, C., et al. (2016). Simulating protection rackets: A case study of the Sicilian mafia. *Autonomous Agents and Multi-Agent Systems*, 1–31

Neumann, M., Frazzica, G., & Punzo, V. (2016). Mechanisms of the embedding of extortion racket systems. In A. Stachowicz-Stanusch, G. Mangia, & A. Caldarelli (Eds.), *Organization in social irresponsibility: Tools and theoretical insights*. Charlotte, NC: Information Age Publishing.

Palazzolo, S., & Prestipino, M. (2007). *Il codice Provenzano*. Laterza.

Paoli, L. (2014). *The Oxford handbook of organized crime*. Oxford: Oxford University Press.

Papageorgiou, E. I., & Salmeron, J. L. (2013). A review of fuzzy cognitive maps research during the last decade. *IEEE Transactions on Fuzzy Systems, 21*(1), 66–79.

Pinotti, P. (2012). *The economic costs of organised crime: Evidence from Southern Italy*. Banca d'Italia. Working papers, number 868.

Prell, C., Hubacek, K., Reed, M., Quinn, C., Jin, N., Holden, J., et al. (2007). If you have a hammer everything looks like a nail: traditional versus participatory model building. *Interdisciplinary Science Reviews, 32*, 263–282.

Punzo, V. (2013a). Le mafie: Struttura organizzativa, dimensione storica, impatto geografico. In M. D'Amato (Ed.), *La mafia allo specchio* (pp. 25–46). Milano: FrancoAngeli.

Punzo, V. (2013b). I protagonisti. Un'analisi qualitativa della rappresentazione del boss Mafioso. In M. D'Amato (Ed.), *La mafia allo specchio* (pp. 158–180). Milano: FrancoAngeli.

Punzo, V. (2016). Un approccio analitico al processo estorsivo: dall'intimidazione alla reazione. In A. La Spina & V. Militello (Eds.), *Dinamiche dell'estorsione e risposte di contrasto tra diritto e società*. Torino: Giappichelli.

Santino, U. (1995). *La mafia interpretata. Dilemmi, stereotipi, paradigmi*. Rubettino.

Scaglione, A. (2008). Il racket delle estorsioni. In A. La Spina (Ed.), *I costi dell'illegalità. Mafia ed estorsioni in Sicilia* (pp. 77–112). Bologna: Il Mulino.

Scaglione, A. (2011). *Le reti mafiose. Cosa Nostra e Camorra: Organizzazioni a confronto*. Milano: FrancoAngeli.

Scaglione, A. (2015). Mafia ed economia. la diffusione del fenomeno estorsivo in Sicilia e i costi dell'illegalità. In A. La Spina, G. Frazzica, V. Punzo, & A. Scaglione (Eds.), *Non è più quella di una volta. La mafia e le attività estorsive in Sicilia*. Soveria Mannelli: Rubbettino.

Sciarrone, R. (2009). *Mafie vecchie, mafie nuove*. Donzelli.

Shapiro, S. S., & Wilk, M. B. (1965). An analysis of variance test for normality (complete samples). *Biometrika, 52*(3–4), 591–611.

Smith, A., & Varese, F. (2001). Payment, protection and punishment: The role of information and reputation in the mafia. *Rationality and Society, 13*, 349–393.

Troitzsch, K. G. (1994). Modeling, simulation and structuralism. In M. Kuokkanen (Ed.), *Idealization VII: Structuralism, idealization and approximation* (pp. 159–177). Amsterdam: Rodopi.

Troitzsch, K. G. (1998). Multilevel process modeling in the social sciences: Mathematical analysis and computer simulation. In A. N. Wim & B. G. Liebrand (Eds.), *Computer modeling of social processes* (pp. 20–36). London: Sage.

Varese, F. (2013). *Mafias on the move: How organized crime conquers new territories*. Princeton, NJ: Princeton University.

Chapter 13
Conclusion

Corinna Elsenbroich, David Anzola, and Nigel Gilbert

This book set out to provide an integrated analysis of the phenomenon of extortion rackets. For this it provided

- A typology of extortion rackets
- A model of the interdependency of state and civil society
- A model of the internal workings of a criminal extortion organisation
- A methodological integration of various data sources on extortion rackets
- Enhanced understanding of extortion rackets as socio-economic systems

We have seen that extortion racketeering is a complex phenomenon, combining criminological and sociological aspects, creating a complex web of interactions between a range of actors, such as the state, civil society and criminal organisations. Understanding of extortion racket dynamics requires an emphasis on the temporal dimension of these forms of criminality, as well as on the institutional arrangements that emerge as adaptive responses, due to the sustained interaction of diverse institutional actors. The second chapter gave a brief example of the diversity in these institutional arrangements worldwide. Traditional mafia-type organisations, for example, all have in common that they position themselves as major social actors in the everyday life of the upperworld, often taking over the provision of several important social goods and services. Symbiotic types of extortion are more common in the zones where these mafias operate, for the role of social brokering performed by these criminal organisations has promoted the generation of social and legal forms of legitimisation, through the development of trust networks and the manipulation of normative frameworks of different social institutions. This situation stands

C. Elsenbroich, Ph.D. (✉) • D. Anzola, Ph.D. • N. Gilbert, Ph.D. Sc.D.
Department of Sociology, Centre for Research in Social Simulation,
University of Surrey, Guildford, Surrey GU2 7XH, UK
e-mail: c.elsenbroich@surrey.ac.uk; d.anzola@surrey.ac.uk; n.gilbert@surrey.ac.uk

C. Elsenbroich et al. (eds.), *Social Dimensions of Organised Crime*,
Computational Social Sciences, DOI 10.1007/978-3-319-45169-5_13

in stark contrast with that of Latin America. A social context characterised by widespread violence, corruption and lack of structural governmental resources to fight organised criminality allows ERSs in the region to often resort to parasitic and predatory forms of extortion, without much care for developing trust networks within society, that eventually allows for social and legal legitimisation of their actions.

The typology presented in Chap. 3, extrapolating from an analysis of diverse global extortion rackets, addressed the most important institutional arrangements that allow for the emergence and perpetuation of systemic extortion. The six categories included in the typology centre on the outcome of sustained temporal interaction of three major institutional actors: the state, civil society and criminal organisation. While the three actors are present in any type of extortion, the typology shows that, in certain social contexts, the actions of one of the three major institutional actors have larger effects on the possible emergence of systemic extortion.

Overall, the typology highlights the role played by spatial, cognitive and normative aspects of extortion rackets as temporal adaptive responses of social systems. Through the awareness of the context specificity of extortion rackets, GLODERS focussed mostly on the third of these aspects: the normative interactions underlying relationships of systemic extortion. Part II addressed the main issues associated with the normative nature of extortion. It, first, discussed social and legal norms and how these norms materialise in the case of extortion rackets. It later provided an empirical example of the changes experienced by the Italian legal framework as an adaptive response to the actions of the mafia over the years.

The models presented in Part III focused on normative relations that have allowed in different ways for extortion in Italy to endure. The chances of reporting or complying, for example, depend on several individual and social aspects that are the product of coordinated interaction between victims and perpetrators. A high level of moderation in the use of violence, both physical and symbolic, has been achieved in extortion rackets in Italy. This is both cause and consequence of the institutional arrangements, a product of the interaction of different social actors. The low level of reports of violence in Italy is partly due to the fact that some extortion in Italy is symbolic.

Part IV honed in on the relevance of social norms, in particular norms engendering trust, for the survival of a criminal organisation. Chapters 10 and 11 provide a novel and interesting view of the rarely seen internal arrangements that allow for the extortion dynamics to extend over time. These chapters show how the relationship of extortion is first destabilised by internal dynamics and then breaks down because of the reaction of the other major social actors, following a disruption of the original normative agreement underlying the extortive relationship.

Previous research on extortion rackets focussed on the criminal organisation as such, through *pentiti* interviews, etc., state responsiveness, through legal analysis, and very specific aspects of the interaction between criminal and civic society in the form of game theory. This book broadens the actors and relationships considered by combining aspects of state behaviour with aspects of civil society (Chaps. 6–9) and looking at the dynamics of a criminal network when internal enforcement ceases and the state enters the scene (Chaps. 10 and 11).

The book presents the integration of a range of data sources and methods. The project comprised a comparative analysis of global extortion rackets, a detailed database of extortion cases in Sicily and Calabria extracted from judicial data, data from international surveys and expert knowledge. Methodologically, GLODERS has two focal points. The first one is the analysis of novel qualitative data sources such as court and police data and newspaper articles and how to make them usable for social research by producing structured data resources. Outputs of this focal point are the detailed database of extortion cases described in Chap. 6 and the conceptual model of a criminal network in Chap. 10. The second focus is in computational modelling. Two models were presented, the first in Chaps. 7 and 8, with two different implementations. It is a model informed by the long history of extortion racketeering in Italy. It looks as the interactions and interdependencies of the different actors in extortion racket systems, focussing on the role of social norms. Social norms are determinants of behaviours and can lead to overwhelming acquiescence with extortion racketeering. The model investigates the potential of social norm change for undermining extortion rackets and the role of the state and NGOs in facilitating norm change. The second model described in Chap. 11 is the result of a detailed case study from a police investigation into the breakup of a money-laundering network. Again, the focus is on the role of norms in sustaining cohesion. This case study exemplifies how small changes can lead to feedback effects on norm change.

The focus on computational modelling informed the collection of new data and facilitated a new use of existing data. Chapter 12 provides a discussion of the calibration and validation of computer models, and the challenges resulting from a lack of data and in particular of the *right kind* of data. The models in this book show in detail how the integration of multiple data sources can solve many of the problems. This ranges from extensive use of expert knowledge to ensure veracity of the dynamics of the models in the face of insufficient longitudinal data to the use of an international attitude survey as an operationalisation of social norms.

In addition to the methodological advances the book deepens the understanding of extortion rackets. The main contribution is the societal embedding of rackets. Looking at the phenomenon from a systemic perspective provides understanding of the interrelationships between top-down and bottom-up effects, e.g. state repression and the change of social norms, and at feedback effects, e.g. when one action triggers the loss of trust throughout the criminal network. One of the most fruitful aspects of GLODERS was the coproduction of the research with stakeholders (cf. Sect. 12.3). The collaboration had an effect on every aspect of the research:

- Stakeholder collaboration allowed access to data not otherwise available, such as the case study of the criminal network.
- Stakeholder collaboration allowed expertise in validating the models and a focus on usability.
- Stakeholder collaboration allowed for dissemination and impact of the research outside of academic circles.

The book presents methodological and substantive advances for the investigation and understanding of extortion rackets. At the same time, it points towards

Index

C. Elsenbroich et al. (eds.), *Social Dimensions of Organised Crime*, Computational Social Sciences, DOI 10.1007/978-3-319-45169-5

Printed in the United States
By Bookmasters